Study Guide to Accompany

MANAGEMENT BY MENU

FOURTH EDITION

LENDAL H. KOTSCHEVAR
DIANE WITHROW

JOHN WILEY & SONS, INC.

Library of Congress Cataloging-in-Publication Data:

ISBN: 978-0470-14053-6

Printed in the United States of America

SKY10023493_122120

CONTENTS

CONTENTS

Preface

Thank you for purchasing this fourth edition of *Management by Menu*. This Study Guide will help you to get more out of each chapter by offering you:

- An outline with which to aid your studies without serving as a substitute for a careful reading of each chapter
- Questions for Review to test your recall of the information contained in the chapter
- Discussion and Activity Questions to test your understanding of the chapter

- Menu Development Activities to assist you in the step by step process of creating a menu for use by actual paying guests
- Menu for Analysis features to allow you to practice concepts introduced in the chapter as you analyze the menus provided in the Study Guide
- An answer key is available at the end of the Study Guide which includes answers to the Questions for Review

Chapter One: A Look Back at the Service Industry

Objectives
1. Describe the historical content that gave birth to and grew the foodservice industry.
2. Identify major contributions made to this growth by industry leaders and innovators.

Outline

Introduction
- The menu is the cornerstone of all food service operations. The development of the menu is both science and art. An understanding of the history of the food service industry and its variety of segments assists each establishment in operating in an effective, proactive manner.

Ancient Food Service
- Food, of course, is necessary for existence and has been prepared and eaten in groups since the beginning of human being. Egyptians had food market places and vendors of prepared foods. Assyrians had the oldest recipe for beer. Chinese travelers had road side inns. Cities had restaurants. Ancient India had laws regulating inns, taverns and food service. Pakistani excavations uncovered restaurant type facilities with stone ovens. Biblical references included cooking contests. Greeks social lives centered on dining including appetizer snacks. Romans feasted extravagantly and brought us *tavernas* from which tavern is derived and from which small community restaurants called *trattorias* evolved. Apicius has long been credited with writing the first known cookbook; *Cookery and Dining in Imperial Rome*. The date of original publication has been the subject of dispute as well as authorship.

Food Service in the Middle Ages
- The fall of the Roman Empire brought erosion to lavish feasting. Hospitality during the Dark Ages was performed as a religious duty by monks in Monasteries. They advanced baking, the concoction of alcoholic beverages and cooking in general. Recipes for food and liqueurs were developed. Food was eaten from wooden trenchers using a dagger and bread as utensils. The French refined eating and a clear pattern of courses emerged with lighter foods served at the start of the meal and heavier at the end with dessert following. Henry VIIIs English court served elaborate foods with less of a course structure than the French.
- Guilds organized food service professionals. *Chaine de Rotissieres* was charted in 12th century Paris. They had a monopoly on the production of their specialties and developed into the classical kitchen organization and codified standards and traditions. The tall chef's hat, the *toque,* became a symbol of the apprentice chef.

1

Early Renaissance-The Development of Haute Cuisine
French Cuisine
- In 1533 Henry II of France married Catherine de Medici of Florence, Italy. She brought master cooks who introduced many great dishes credited to France. She taught the court to eat with utensils. These were then carried with the diner when dining away from home. French courts continued and expanded on the ascendancy of *haute cuisine*. Sauces were named after nobility hosts. Mistresses of Louis XV were excellent cooks. Menus were elaborate with 20 or more dishes served in three courses though simplification won out in time.
- End of French revolution did not end love of dining. Servants now cooked in restaurants instead of homes. Gourmet writers of this period included Brillat-Savarin (*The Physiology of Taste*), Dumas pere (*The Grand Dictionaire de Cuisine*) and Grimrod de la Reyniere, editor of the world's first gourmet magazine.

The Coming of the Restaurant
- Common folk ate from crude inns and taverns away from home but places to dine out that were not private homes did not exist. Prisons and hospitals served only the most basic minimum fare.
- In 1600 the first coffeehouses (cafes) appeared and spread rapidly. They served mostly coffee, cocoa, wine and not much food. They were the forerunner of the modern restaurant.
- In 1760 Boulanger opened a place that served soups that were believed to restore health. These soups were called *restaurers* and he called his establishment a *restorante,* now restaurant.
- Guilds opposed Boulanger's infringement and sued. Through deft politicking, Boulanger won the right to operate and expanded his menu. Coffeehouses followed the example and became restaurants.

Other National Cuisines
- England had worthy tables with food and nobility arriving to its seaports. Countries represented in the blend were Spain and Portugal. Sponge cake, Spanish hams and bacon, sherry and port incorporated into the British diet.
- Russia developed a fine and original cuisine, partially through importing French chefs to royal kitchens to work with local products. Gatherings with vodka featured appetizers brought on platters held high. Russian dishes include stroganoff, caviar, borscht and vodka.
- Italy's distinctive cuisine is dominated by pastas though regional dishes based on what was produced locally distinguish various aspects of Italian cuisine.

The Industrial Revolution
- Economic and societal changes at the end of the 18th century brought changes to commerce. The middle class emerged and demanded a high food standard served in clubs. Lower middle class also dined out as it became affordable.

The Advancement of Science

- Science defined by philosophy gave way to the inductive method, advancing technology and the standard of living including food processing.
- Appert discovered canning, allowing food to be preserved and stored. This enlarged the year round food supply and appeared to end mass starvation.

The Golden Age of Cuisine

Careme

- Golden Age of Cuisine began around 1800 with Careme and ended with Escoffier, both remarkably talented chefs. Careme was apprenticed to a restaurant owner and originally trained as a pastry chef. He branched out and left behind his early architectural ambitions.
- Careme developed the progression of courses and sequencing of accompanying wines. He perfected *consommé*, sauces and highly decorated foods such as ice carvings. He trained many famous chefs and wrote of his ideas of food, leaving a legacy that has influenced culinarians throughout history.

Escoffier and Ritz

- Escoffier was an innovator of fine foods like Careme. He perfected the classical organization of kitchen workers defining the responsibility of each and decorum of all.
- He introduced the *aboyeur,* who announced servers' orders to kitchen workers.
- He wrote articles and books and invented dishes though the sound uses of scientifically based quantity food production techniques.
- Escoffier simplified the menu from the number of food served at each course, following Careme's lead on the progression of courses in a formal meal from light to heavy.
- He teamed up with hotelier Cesar Ritz to operate many of Europe's finest hotels raising the level of dining and staying away from home to high peaks.

Food Service in the United States

The Early Years

Institutional Feeding

- Orphanages, hospitals and prisons used fireplaces or beds of coals. Ground cooking involved cooking in an oven in a hole with food cooked in stored heat. Universities had family style dining or table service in rooms called *commons*. Others had private apartment style kitchens where a cook-slave prepared food.

Hotels Appear

- 1818-1850 saw enormous growth in numbers of hotels to meet demand in cities such as New York and Chicago. Famous chefs were hired to provide food. Famous examples include the Astor House, Palmer House, Brown Place, and the Butler. Western hotels flourished with newly rich from the Gold Rush to California.

- Railroads built after the Civil War brought small hotels near railway stations and included nearby restaurants.

Dining Trends

- Lorenzo Delmonico of New York reflected the elegance and luxury of hotels in his Delmonico's restaurants, one of the first restaurant "chains".
- Workers left home to toil in factories and offices. Coffee shops and restaurants provided modest lunch at prices suited to these employees. *Nickelodeon* automatic food dispensers were developed by Horn and Hardart. Mobile food units and facility kitchen and dining rooms were also used for employee meals.
- Disposable income rose as did mobility spurred by the invention of the automobile. Electricity powered refrigerators, freezers, mixers and dishwashers, preserving food and labor resources.

Postwar Expansion

- Post WWII, foodservice grew rapidly due to the expansion of industrial feeding. 1946's National School Lunch Act started educational food service on a massive scale. Dieticians were sought out for institutional as well as commercial enterprises. Cornell University introduced the first hotel school in the US which added restaurant, institutional and tourism curriculum.
- Statler led the way in development of vast hotel chains. People ate out to fulfill social and psychological desires.

Quick Service and Corporate Concepts

- Fast Food or Quick Service served food that could be prepared and eaten quickly, starting with the White Castle hamburger units of the 1930s. Chicken was also a popular entrée item.
- Corporate Concepts or Chains began in conjunction with QSRs. Assembly production of these lower priced, lower margin items was made profitable through volume sales. Today, QSR chains make up a very significant portion of restaurants and an even higher proportion of sales nationally. Corporations gained on the predominance of independently owned entrepreneurships.
- Take out and curbside service is important and is expected to grow. Dual incomes, busy lives and convenience motivate consumers. This segment includes QSR drive-throughs, grocery delis, and cafeterias. Curbside service is offered by casual and fine dining establishments in regular or family size portions. Close to half of the food sold by food services is considered take-out. Dave Thomas was a pioneer of "adult" quick service with his Wendy's concept including a salad and baked potato bar. He created the modern pick-up window.
- Rich Melman's Lettuce Entertain You Enterprises is renowned for remarkable success at creating over thirty separate menu concepts. Listening to people and giving them what they want before they know they want it with a sense of humor and style have been guiding principles to this creative empire.

Continual Growth
- The foodservice industry has grown tremendously with no signs of slowing. It is predicted people will soon be eating more than half of meals outside the home.

Questions for Review

1. What is the point in studying the history of food service?
   a. In order to better understand our mistakes
   b. In order to explain how to manage all restaurants
   c. In order to successfully respond to conditions today and better predict tomorrow
   d. To pay tribute to those who blazed the trail for today
2. How long have people been preparing food in and for groups?
   a. For the last 150 years
   b. Since ancient times- 10,000 BC tribes in Denmark and Scotland
   c. Since the Middle Ages when monks prepared food in monasteries
   d. Since the Renaissance when chefs prepared food for royal courts
3. Who created the first recorded recipe for Beer?
   a. Egyptian Pharaoh Tutankhamen
   b. Franciscan Monks
   c. Madame du Barry
   d. Assyrians
4. Which of the following was NOT associated with ancient Greek food service?
   a. Emperor Lucullus who loved lavish banquets
   b. Bacchus, the God of Wine
   c. Epicurus who spread the philosophy of good eating
   d. Professional cooks who could copy write their recipes and were well respected in society.
5. How are *Trattorias* associated with Roman food service?
   a. They are Pigs feet, kept at taverns in earthen ware jars
   b. They were forerunners to the "all you can eat" buffet
   c. Small community restaurants in modern Italy derived from ancient taverns
   d. Inspectors charged with seeing that food service operations were complying with the law
6. What is the first known cookbook?
   a. The Joy of Cooking
   b. Cookery and Dining in Imperial Rome
   c. The Grand Dictionaire de Cuisine
   d. Golden Toque Recipets de Gastronimique
7. How did Apicius, credited author of the first known cookbook, meet his death?
   a. Tragically choking on a bone mistakenly left in boneless breast of duck
   b. Hardening of the arteries, resulting in heart failure
   c. Peacefully in bed with a snifter of fine brandy, surrounded by friends
   d. Sadly, he committed suicide when a lavish banquet bankrupted him

8. What was a trencher, used in the Middle Ages?
   a. An early bus tub
   b. A cooking vessel made of earthenware
   c. A fire pit dug for cooking large pieces of meat slowly
   d. A crude wooden dish for eating

9. The black chefs hat once symbolized:
   a. The mark of a master chef, nominated by his peers to wear it
   b. The shameful warning that a chef had caused a food related death
   c. The sign of apprenticeship entering the final year
   d. Particular skill grilling meats

10. Henry II of France is remarkable in our history because:
    a. He preferred to eat his food with a dagger and a piece of bread
    b. He inspired the dish *Potage Henri*, which is ladled into a large tureen and has pieces of chicken and beef in it
    c. He married Catherine de Medici of Florence, Italy who brought her own cooks to the castle and launched *haute cuisine*
    d. He was very active in the development of good schools for training chefs and cooks

11. The founder of what is thought of as the modern restaurant was:
    a. Catherine of Aragon
    b. Boulanger
    c. Escoffier
    d. Ray Kroc

12. The *Chaine de Rotissieres* and *Traiteurs* were:
    a. The first organized chef unions
    b. Guilds of professionals with exclusive rights to their trade
    c. Secret chef gangs committing recipe theft on a grand, organized scale
    d. Exclusive social clubs for wealthy gourmets

13. Which of the following food or beverage is of Spanish origin but now associated with England?
    a. The Bloody Mary
    b. Sausages or "bangers"
    c. Sherry
    d. Steak and Kidney Pie

14. The Industrial Revolution was most important to the history of food service in what manner?
    a. It strengthened the Guild system, ensuring professionalism in food preparation
    b. Dining out became more popular because newly wealthy or middle class could afford and desired well prepared food.
    c. Assembly line methods of food preparation increased labor efficiency and paved the way for quick service restaurants.
    d. The invention of the automobile spurred demand for drive throughs and curb side service.

15. Who developed the basic progression of courses and trained a large number of famous chefs?
    a. Marie-Antoine Careme
    b. Marcel Escoffier
    c. Dave Thomas
    d. Rich Melman

Discussion and Activity Questions

1. Examine www.leye.com , Lettuce Entertain You Enterprise's web site. What ideas do you take from this website? Which restaurant(s) would you make a reservation for if you:
    a. Wanted a romantic evening to charm the number one love interest in your life?
    b. Wanted a business dinner featuring good food in a relaxed atmosphere that would appeal to any taste.
    c. Wanted a get together evening with friends who have adventuresome palates but not big budgets for eating out.
    d. A celebration with the same friends upon coming into a good deal of money.
    e. A party to celebrate your college graduation
    f. A family reunion
    g. A High School reunion
    h. Wanted to impress and honor your supervisor (or professor)
    i. Wanted a food adventure that appeals to you
    j. You are in Las Vegas and want to spend some of your casino winnings
    k. Just plain hungry and in a hurry

    Explain your choices and describe what you would order

2. What does history have to teach us that can be applied to food service today?

3. Strictly using food service as the criteria, which period of time would you like to have lived in and what food service part would you like to have been involved in playing? Explain the reasoning for your answer. Could your choice be applied to today's world?

4. What was so significant about having recipes? The canning process? Coffeehouses?

5. Explain the contributions made by Rich Melman and Dave Thomas to food service. What contributions would you like to make to our industry?

6. What is the Slow Food Movement and why did it come about?

Menu Development Activity

This activity will continue throughout the manual. You will want to save your work product to use in future activities.

Brainstorm concepts for a restaurant or food service operation that you would like to own or run someday. To brainstorm is to write down every idea, no matter how unconventional or impractical, and worry about critiquing or filtering later. You want a free flow of ideas, uninhibited, at this stage, by practicalities. Write them down. If assigned, target specific markets for these ideas such as: Elementary School Kids and their Parents, College Students, Academic faculty and staff, Retirees, Vacationers etc.

Menus for Analysis

The following menus are adapted from actual menus in the NYC Public Library collection. Alter the menu to fit today's customers. How extensive are your alterations? Were there items that you are not familiar with? Explain why you omitted items that you did.

Washington's Birthday 1892
Sponsored by the Sons of the American Revolution and held at the Hotel Brunswick, NYC

Blue Point Oysters
Hors D'Oeuvre
Celery   Olives   Radishes
Caviar

Potage
Clear Green Turtle

Poisson
Lobster Newburg

Entrées
Escaloppes of Filet Pelissier   Duchesse Potatoes
String Beans
Sweetbreads   French Peas

Roman Punch

Roti

Golden Plovers on Toast
Salad

Entremets Sucres
Fancy Forms Ice Cream   Jelly
Cream Puffs   Petit Fours

Dessert

Roquefort   Camembert   Fruit
Coffee
Champagne

Menu to Honor John Francis Maguire
Delmonico's NYC, NY March 14, 1867

Menu

Soup
Asparagus Cream

Removes
*Timbale d'Ecarlate

Entrées
Filet of Beef with Mushrooms
Boiled Capons with Celery Sauce
Breast of Canvas Back (duck) with Olives

Roman Punch

Saddle of Mutton with Stuffed Lettuce
Roast Quails
Chicory Salad

Dessert

English Plum Pudding
Jellies Bavaroise: Assorted Cakes
Fancy Pieces
Charlotte Russe

(*vintage tomato molded salad)

Wine

Chablis
Amontillo Sherry
Roederer Champagne
Chateau Perganson
Whiskey

Chapter Two: Profile of the Modern Foodservice Industry

Objectives
1. Characterize the distinguishing features of various types of foodservice operations and describe situations where categories may blur.
2. Differentiate between the following major segments of the foodservice industry: commercial, institutional, transportation, health services, clubs, military feeding and central commissaries.
3. Identify economic, social, labor, health and technological trends that are likely to impact the future of the foodservice industry.
4. Speculate on industry segments that may grow or recede in popularity and provide justification for predictions.

Outline

Introduction
- The foodservice industry consists of many diverse operations that provide food and drink for people away from home. This huge industry hires more people than any other in the world. It is the US largest private sector employer and one of the most consistent growth industries, particularly after WWII.

A Diverse Industry
- Units differ in food and drink offered, preparation methods and service as well as customers, merchandising and operations. Eating and drinking establishments make up the most with nearly 70% followed by; Managed Services, Lodging Places, Retail-hosts, Recreation and sports, and all others.

Income and Costs
- Food, beverage and labor costs represent the prime costs of operation with 65% regarded as a goal in order to achieve profitability. As labor costs rise, operators are increasingly challenged.

Segments of the Foodservice Industry
- Menu planner must know the characteristics of the operation and the guest it is seeking to serve or the menu and thus operation will fail.

    Commercial Operations
    - This is the largest division, responsible for most of the sales and includes any operation that sells food and beverages for profit.

        Restaurants and Lunch Rooms
        - This segment is largest representing 28% of commercial units. They range from fine dining or "white tablecloth" serving lunch, dinner and alcohol to Lunch rooms, so named due to the mid day being their primary business. They may also serve breakfast and dinner and are simple in nature. Some serve alcohol.

- Alcohol consumption has gone down and responsibility for serving alcohol responsibly has increased. Emphasis on quality versus quantity has been one strategy.

Family Restaurants
- Cater to a more casual customer group and *may* not serve alcohol
- Moderately priced with higher turnover than fine dining, these generally offer a children's menu
- Cafeterias, buffets, and upscale quick service restaurants may be included along with seated service.

California-menu Food Services
- Originating in that state, the menu offers breakfast, lunch and dinner items served at all hours of the day
- Often opened 24 hours per day

Limited-menu (Quick Service) Restaurants
- Quick service have a high turnover rate and offer simple fare
- Drive through and take out business are significant for this segment
- Many are part of a chain or franchise. This segment is growing including ethnic and healthy offerings often desired by mature consumers. Pizza, burgers, chicken, sandwiches, ice cream and snacks remain staples of this segment.

Commercial Cafeterias
- Guests select steam table items priced individually, per plate or per ounce and then take the plate to a table to eat.

Social Caterers
- Caterers prepare food and beverages in a central kitchen often to take elsewhere for service
- Commercial restaurants and grocers are competition as many have added this service to expand their business
- Events range from guest pick up of prepared dishes to provision of everything needed for a formal event including entertainment with parties from very small to thousands

Frozen Dessert Units
- These popular units are often part of a chain or franchise
- Offerings range from ice cream, soft serve, smoothies or frozen custard only to things that go with frozen deserts. Frozen dessert specialty items are often offered that may be custom decorated or generically for take away.

Bars and Taverns

- Free standing not including those operated in conjunction with another type of food service
- Food is generally limited menu suitable to accompany alcoholic beverages
- Reputation is for social gathering place with recreation such as television, pool tables, video games etc.
- Due to liquor liability laws, a shift has been seen toward emphasizing family fun versus quantity drinking

Managed Services

- Food contractors are paid to take over the food service for pay, usually because the main purpose of the enterprise is not food service but there is an ancillary need for it.
- Customers include; manufacturing, industrial plants, hospitals, schools, colleges, universities, airlines and sports centers
- Contractors may be fully subsidized or only partially depending on the nature of the operation
- Cafeterias, Executive dining rooms, breaks, vending, coffee shops and quick service units as well as catering are operating venues
- Transportation catering in the form of free meals on airlines has all but gone from domestic flights under three hours. When food is offered, it may be for sale in limited quantities. International and cross continental flights continue to offer meal service.

Lodging Places

- Food and Beverage Department in a hotel may offer; coffee shops, banquet facilities, snack bars, bars, a nightclub, fine dining rooms, specialty restaurants, room service, a cafeteria and possibly an employee dining room
- Employees may include; Food and Beverage Director, Executive Chef, Maitre d'hotel, host(ess), Banqueting or Catering director, and Steward as well as the traditional cooks, servers, bussers and dishwashers
- Luxury foods and fine wines as well as simpler fare may be offered
- Many lodging units cater to conventions, parties, meetings, receptions and other occasions. Designated kitchen and personnel may be required for these.
- Attracting in house customers as well as outside trade is a challenge met through creative menus and promotions

Retail Hosts

- Establishments "hosting" food service units include; drug stores, department, variety, grocery stores, gas stations and "mega" stores
- Offerings vary according to the needs of the shoppers frequenting the host stores.
- Grocery and Supermarket bakeries, delis and catering are competitors with food service operators

Recreational Food Service
- This field consists of sports stadiums, coliseums, arenas, convention centers, concert facilities, movies, bowling lanes, amusement parks, racetracks, expositions, carnivals, circuses, zoos, gardens, parks, and roller and ice skating rinks.
- Offerings range from casual concession food such as hot dogs and beer to elaborate meals
- Food is sold by circulating personnel or at counters or served in dining areas

Mobile Caterers
- Trucks, vans and mobile units deliver to factories, construction sites and other work and play places
- Work days can last 10-12 hours with up to 40 or 50 stops over 15-20 miles operating when demand is present
- Food must be acceptable, moderately priced and sanitary. Typically units offer similar food to vending operations.

Vending
- Growing in popular, vending services; schools, industrial plants, office complexes, retail stores, hospitals, nursing homes, colleges and universities.
- Offerings include; sandwiches, pastries, dairy snacks, casseroles, platters and salads. Central commissaries produce 500-600 meals a day to be profitable, otherwise food is purchased prepared.
- 20% of in-plant feeding is completely vended products. Popular foods served in sanitary fashion are important for success.
- Substantial investment is required for machines and transport vehicles. Simple machines start at $3,000.00 with a five year period required to recover costs

Institutional Feeding
- This segment consists mainly of noncommercial and commercial private and public organizations that operate food services in support of the actual purpose of the establishment. This service is increasingly performed by managed services with the same clientele.

Employee Feeding
- Industrial plants, office buildings, public school systems, colleges, universities, and others are primarily serviced by Managed Services. Seagoing ships and inland waterway vessels are also included.
- These popular programs include all meal periods with healthful foods gaining in popularity as in the commercial sector. Venues include; cafeterias, vending and executive dining rooms.

Educational Feeding

Elementary and Secondary Schools

- Primary and Secondary Managed Service feeding was projected to grow faster than any other industry segment in 2006
- National School Lunch Program of 1946 is administered by the FDA to offer a market for agricultural products and to serve nutritional foods to children at low or no cost and includes the Special Milk Program
- Schools must comply with a school lunch meal pattern specified by the federal government which specifies healthy foods. After school snacks for children under 15 are included.
- Food must be popular with children who can be particular. To date, more than 187 billion lunches have been served.

Colleges and Universities

- Venues include residence halls, student unions and other food services
- Meal plans may include all or fewer than all meals. Technology is used to swipe prepaid cards for residence halls and even off campus students
- Techniques besides the traditional cafeteria line include scramble stations, alternative venues, snack bars, coffeehouses, faculty clubs and dining rooms suited to student tastes

Transportation Feeding

Airline Feeding

- Increasingly done only on international flights and cross continental flights of more than 3 hours
- Much like a central commissary with rapid system for unloading used items and loading on new items which have been produced efficiently, freshly and with sanitary methods
- Six day cycle menus are used to satisfy frequent fliers though some do not change with several menus for first class, coach and special diets offered on an airlines own dishes, if applicable
- Considerable storage and good flexibility is required of central commissary type operations
- Airport venues offering box meals and snacks to passengers may offer new opportunities

Railroad and Bus Food Services

- High quality food prepared on board served specially may be limited to Luxury ride trains. Snack foods and sandwiches are more typical

- Bus terminals may have quick service operations and remains popular. Food is rarely sold on the bus.

Health Services Feeding

Hospitals
- Nutritional needs of patients generally require the services of a Nutritionist or Dietician to direct
- General menus are modified for those on restricted diets using a cycle menu
- Hospitals use either centralized or decentralized services where prepared quick chill or convenience foods are heated through transportation devices or microwave or conventional ovens
- Hospital food has long been the source of complaints from patients who, under the circumstances may not like any food.
- Innovations such as room service, special menus and promotions are used to increase acceptance

Long-term Health Facilities
- Much like hospitals with the addition of elderly patients who may need soft diets.
- Traditionally, budgets in these facilities have been low but with an increasingly aging population with a range of income levels, opportunities have increased in this field

Clubs
- Members and guests who wish to enjoy luxury or casual food in dining rooms, snack bars, and catering venues comprise the bulk of this segment. Country and City clubs are included
- In order to operate within budget, members are sometimes charged a monthly rate for food

Military Feeding
- Changes continue and include changes in troops to more married enlistees needing family centers with post exchanges that offer quick service and snack bars
- Feeding troops is still a priority with more and healthier choices

Central Commissaries
- Chains, large operations and also school systems use central commissaries to produce much of the food they then ship to satellite units where they are reprocessed and served
- Central commissaries may resemble factories in which foods are mass produced on assembly lines using special equipment
- Some chains have shut down units when cost reductions are not significant but school systems and other institutions continue to use them.

The Future of Food Service
- The National Restaurant Association puts out an annual Restaurant Industry Forecast. This seeks to predict what may occur in the year to come

Forces of Change

The Economy
- The food service industry will continue to grow in a fiercely competitive environment
- The commercial segment of the industry is very dependent on disposable income. This may slow down or increase if wages remain stagnant or actually decrease when adjusted for inflation.

The Social Pattern
- Birth rates have slowed as well as the number of young people while the number of those over 65 continues to increase.
- Married couples with children have decreased to a quarter of the population with single people with children increasing. A significant increase has occurred in the number of people living alone.
- Ethnic differences including the rapid increase in the number of Hispanics as a percentage of population will also affect food service offerings. Emphasis on healthful foods will also affect menu development.

The Labor Force
- The "white" population is decreasing as a percentage with Africans increasing rapidly and Asians and Hispanics more so- expected to make up 8% and nearly 24% of the population by 2050 respectively.
- Consequently, the faces of employees in our industry will reflect these changes with women making up over 55% of the labor force.
- Training needs will increase with Hospitality Schools needing to fine tune programs for future managers.
- The DOL has identified Hospitality as a high growth industry with the need to counter negative stereotypes, expand the youth labor pool and target untapped labor pools. Reducing turnover and addressing language skills with consistent training and certifications were also cited.

Healthful Foods
- Customers today demand healthful foods
- Culinology is the blend of culinary arts and food science. This field offers opportunity to produce foods that enhance health without sacrificing taste. The Research Chefs Association promotes Culinology.

Recruitment and Retention
- Food service has more people on the payroll than any other- 40% of all Americans will work in the industry with 27% as their first job.

- Often a temporary job in a career leading elsewhere. Low wages and irregular hours also contribute to high turnover.
- Turnover is expensive. To combat it, image, opportunity for advancement and good management incomes are publicized.

Government Regulations
- Once ignored, the food service industry has been singled out for special regulations beyond general business laws
- Specific laws include; truth-in-menu, nonsmoking areas, sanitation, 3rd party liability and liquor laws written at the state and local levels
- Federal laws include; tip reporting and documentation of legal status to work

Foodservice Industry Trends
- Trend watching is important to facilitate growth. These include changes in preferences, menu trends and technology in food, equipment and supplies
- It is important to learn to differentiate between a trend and a fad and to apply it to local tastes and eating patterns
- Convenience, cost reduction, demand for healthful food, takeout, curbside and delivery are all current trends.

Technology
- Many changes have occurred in production and service including sanitation and what foods are available on the market
- Equipment for food preparation, energy efficient appliances, and innovation at the supplier level including production and packaging have advanced quality
- Labeling requirements, new fat substitutes and non caloric sweeteners advance healthful selections
- Computer uses include operations, internet marketing and more.

Questions for Review
1. Which of the following is a commercial operation?
   a. Hospital
   b. Prison
   c. Vending
   d. Elementary School
2. An operation with a California style menu:
   a. Has all you can eat health food
   b. Offers only limited menu items
   c. Prepares food in a central kitchen and delivers to satellite facilities
   d. Offers breakfasts, lunches, dinners, snacks and other foods at any hour, all on one menu

3. A gas station that has a submarine sandwich store attached to it is a type of:
   a. Transportation Feeding
   b. Social Caterer
   c. Retail Host
   d. Quick Service
4. Menu items that have grown in popularity include:
   a. Healthy choices that are lower in fat and sugar
   b. Deep fried foods
   c. The cuisine of Patagonia
   d. Soft, bland items that have low salt content
5. The sort of food service whereby service personnel circulate through the stands, vending their wares is:
   a. Recreation Food Service
   b. Concessionaires
   c. Ball Park Hawker
   d. Independent Contractor
6. One of the fastest growing segments of our industry, providing food service to such operations as hospitals, elementary schools and industry for a price is:
   a. Institutional Caterers
   b. Managed Services
   c. Health Care Concessions
   d. Mobile Caterers
7. When operations whose primary business is not food service provides it themselves, this is referred to as:
   a. Institutional Food Service
   b. Self-Catered
   c. Independent Contractor
   d. Franchisee
8. An operation whose employees include a Food and Beverage Director, Executive Chef, Catering or Banquet Manager and Steward is classified as:
   a. Large scale catering
   b. Lodging Places
   c. Fine Dining
   d. Mobile Caterer
9. Frozen Dessert Units are characterized by:
   a. Jauntily painted trucks which play tinkling music to attract children
   b. Large play areas and animal or cartoon characters in costumes, performing
   c. Self contained carts operated by a single vendor in fine weather
   d. Menus that may include things that go with dessert, ability to order frozen dessert special occasion items and possible chain or franchise affiliation

10. Airline Feeding:
    a. Is a growth segment of our industry and point of differentiation amongst carriers
    b. Is characterized by the distribution of tiny bags of pretzels and overpriced beverages to a traveling public shoehorned into tightly packed seats
    c. Is now generally limited to continental flights over three hours and international travel
    d. Remains one of the last bastions of culinary excellence
11. Bars and Taverns have redefined themselves as places for:
    a. Pickled pigs feet, slim jims and hard boiled eggs in a jar
    b. Those who love alcohol to let themselves drop inhibitions and "get loose"
    c. Fine wine, coffee beverages and cigars
    d. Family fun with entertainment and popular snack food and more
12. Which of the following statements about lunch rooms is *false*?
    a. Lunch rooms generally serve breakfast and dinner as well as the mid-day meal
    b. Lunch rooms may serve alcohol
    c. Lunch rooms offer complete meals served at a leisurely pace
    d. Shoppers, office workers and those traveling in high traffic areas make up the bulk of their trade
13. A segment in which drive through business has experienced rapid growth, a maturation of the market has provided challenges and limited menus, prepared efficiently are often served by a chain or franchise affiliated operation is called:
    a. Employee Feeding
    b. Quick Service
    c. Recreational Food Service
    d. McDonald's
14. Dram shop or third party liquor liability refers to:
    a. The responsibility of the guest to drink responsibly
    b. The legal responsibility of the party profiting from the sale of alcohol to serve it responsibly
    c. The responsibility of the purveyors of liquor to sell safe products to food service operations
    d. The requirement that all establishments that serve alcohol, serve coffee and other caffeinated beverages to assist in patron sobriety
15. Technological advancements in food service include all of the following *except:*
    a. Aquaculture
    b. Convection Cooking
    c. Assembly line production
    d. Hydroponics

Discussion and Activity Questions

1. Using the web to learn more about the menus and operations:
- Classify the following according to the industry segments discussed in this chapter
- Examine a menu or meal plan and describe any features that particularly suit the intended market
   a. Sodexho Marriott
   b. Cold Stone Creamery
   c. Moe's Southwest Grill
   d. Angus Barn, Raleigh, NC
   e. Sonic
   f. Ritz-Carlton, Chicago
   g. American Airlines

2. Discuss your own ideas for mobile catering. Have fun. What items would you offer for sale? What impressions do you have of that segment of the industry?

3. Are there any new segments that you imagine might emerge in the future? What might they be? Are there any segments that you perceive may fade away or undergo major changes?

4. What is the value in classifying food service industry segments?

5. What are additional technological advances that you perceive may assist our industry?

6. How do changes in the demographics and family structure of a country effect food service? Be specific.

Menu Development Activity

Take the concept(s) that you brainstormed in the first chapter and classify them according to segment or segments represented. Does classification add anything further to fleshing out your concept(s)?

Examine some menus illustrating successful concepts in the classifications you have identified. What new thoughts or directions do these lead to?

Menus for Analysis

Following are two menus. Analyze each for the following:

- What industry segment does this menu represent?
- What food service trends are addressed by the menu?
- Would this menu be successful in your community? Where?

Good Morning

Beverages

Freshly Brewed Coffee from just ground beans

Florida Orange Juice

Tomato Juice
> With a splash of tabasco

Bottled Water

Lighter Appetite

Organic Vanilla Yogurt
> Served with Fresh Bananas with Toasted Granola

NY Style Bagel with Philadelphia Style Cream Cheese
> Plain, Cinnamon or Multi Grain

Hickory Smoked Salmon Bagel
> with pickled red onion and cucumber

Country Ham, Sausage or Applewood Bacon Biscuit

Breakfast Sandwich
> Choice of meat with egg and cheese on country biscuit

Breakfast

Unless specified, choice of Hash Browns, or Grits and Toast or Country Biscuit

Omelet
> with cheese and your choice of breakfast meat

Platter
> Eggs any style with your choice of breakfast meat

Shrimp and Grits
> Smothered shrimp in low country gravy with stoneground grits

Fish and Grits
> Fried Flounder, Grape Tomatoes and Stone Ground Grits

Vegetarian Breakfast Wrap
> Scrambled Eggs, Navajo Black Beans, Mango Salsa and Jack Cheese

Sides

Stoneground Grits

Cheese Grits

Hashbrowns

Toast or Biscuit

Big Boy's Sandwich Grill

Delivery to the Gang

Great for working lunches, company picnic or teams on the go- We deliver to you and set up, included in the price

$6.99 per person: Includes: One half sandwich, chips, pickle, cookie, soft drink, condiments and disposable ware

$5.99 per person: Includes: One half sandwich, choice of chips, pickle or cookie, soft drink condiments and disposable ware

$4.99 per person: Includes: One half sandwich, your choice of chips, pickle or cookie, condiments and disposable ware

Catering

Please call ahead so we will be ready for you

Tossed Salad Tray

House blend of greens with red onions, olives, peppers, tomatoes, cucumbers and carrots and your choice of dressing

Hearty Salad Tray

Centerpiece of your choice of Chicken, Tuna or Egg Salad with grilled bread and garden fixing's

Sandwich Tray

Choice of any three sandwich selections served by the half with a centerpiece of condiments and pickles

Cookie Tray

Fresh baked chocolate chip cookies, gooey with chunks of chocolate, Cranberry Oatmeal and Old Fashioned Peanut Butter

Sandwiches

TLT Turkey, Lettuce, Tomato and Mayo
Deli Ham, Swiss Cheese, Special Mustard
Muffaletta Ham, Salami, Provolone, Olives
Roast Beef Roast Beef, Onion, Cheese
Vegetarian Swiss and Provolone Cheeses, Lettuce, Italian Dressing, Tomato, Pickle
Salad Sandwich Choice of Chicken, Tuna or Egg Salad with Lettuce and Tomato

Half Sandwich 3.75, Whole 4.75 Big Boy 5.75

Hot Sandwiches

Grilled Chicken Chicken and Swiss or grilled vegetables
Pulled Pork Served with cole slaw, in low country fashion- vinegar based BBQ

Half Sandwich 3.75, Whole 4.75 Big Boy 5.75

Fresh Tossed Salads

House Mixed greens with choice of dressing
4.00
Grilled Chicken Add Freshly Grilled boneless breast
6.25

Soups

Served in fresh, homemade edible bread bowls

Seafood Chowder
Cream of Broccoli
Chicken Noodle
Garden Vegetable

4.25

Chapter Three: Planning a Menu

Objectives
1. Identify and characterize various menus used in the foodservice industry and explain the needs met by each variety.
2. Describe what is meant by "Meal Plan" and explain how menus are developed for them.
3. Explain how menus are organized and structured traditionally and the process by which they are derived.
4. Describe the various tools used to plan menus.
5. Compare and contrast institutional and commercial menus.

Outline

Introduction
- For consumers, a menu is a list of offerings, for operators it is a strategic document that defines the purpose of the establishment and every phase of the operation. It is first a working document used by managers and second a published announcement of what is offered to patrons. A good menu should lead patrons to selections that satisfy both the guest and the operator. Designing a menu that sells is critical both for commercial and noncommercial operations. Regardless of the status of the diner, the menu should be an invitation to select something that pleases.

What is a Menu?
- Originally, offerings were recited by the server. Sign boards near the entrance with Maitre'd describing offerings was customary. Later, small sign boards were hung on waiters' belts and as menus became more complex, written menus entered into general use. Later, bills of fare were distributed to guests to take home as souvenirs and advertisements. Menu means "a detailed list".

  The Purpose of a Menu
  - o To inform- what is available at what price (and how to produce it)
  - o Menu is the basis for all other planning
  - o A management team must first create a Mission Statement, and then develop the Financial Plan. Related to this is the establishment of operating budgets stated on an annual basis in monthly increments.

  Who Prepares the Menu?
  - o Menu Planners must be skilled in: knowledge of operation and market including extensive understanding of foods and preparation, operational constraints, graphic visualization, and communication.
  - o Working in groups may compensate for gaps in knowledge. Front of House personnel understand guests preferences, Back of House know the kitchen and have skills needed to create selections.
  - o Menu planning is the most critical step in defining the operation

Tools Needed for Menu Planning
- o Quiet workplace with space to spread out
- o Records on past menu performance, ideas, sales mix data, special occasion and holiday menus, and costs and seasonality of potential menu items

Market Research
- o Link of information concerning consuming public's buying preferences and how well seller's business meets those preferences *as viewed by the potential buyer*.
- o Formal research is scientific process of data collection, analysis, and communication of findings
- o More specific information is most valuable but may be expensive to obtain. Sources of information include: Market Research Firms, Federal Government, Financial Newspapers and Periodicals, Stock and Investment firms, and Trade Associations such as the NRA and its state branches.
- o Studies can be conducted by marketing firms (expensive), or, more feasibly, a college might perform one, advertising agencies, or an operation could perform their own research. Local Convention and Visitor's Bureaus and Chambers of Commerce often collect useful data.
- o First, it must be determined what kind of information is needed, next how the research will be conducted: interviews, questionnaires, observation and focus groups are all methods used with open or close ended questions.

Menu Planning Factors
- • New menu should be planned to allow for delivery of items, schedule labor and for printing. Lead time could be as long as six months.

Number of Menus
- o A single operation may need a number of menus depending on the types of enterprises engaged; number and types of outlets, events, and season.

Types of Menus

À la carte Menu
- o Offers food items separately at a separate price, allowing customer to "build" meal to suit.
- o Often leads to increased check averages and profits if waste is kept at a minimum

Table d'hôte Menu
- o Several food items are grouped together at a single price with choice sometimes offered
- o This type of menu appeals to patrons who are unfamiliar with cuisine.
- o This limits the number of entrées which must be produced
- o À la carte and table d'hôte menus are often combined

Du Jour Menu
- o Group of food items served only for *that day* (du jour) often associated with daily special such as soup du jour
- o Often combined with à la carte and table d'hôte items
- o Frequently allows utilization of overstocked or near expiration date products so that they become profit and not loss

Limited Menu
- o Fewer choices are offered to simplify operations and keep costs down

Cycle Menu
- o A cycle menu refers to several menus that are offered in rotation. The purpose of the cycle menu is to inject variety into an operation catering to "captive" patronage in a manner that effectively and efficiently utilizes food product. The length of the cycle depends on the average "stay" of the patron.

California Menu
- o Originating in that state, this menu offers breakfast, snack, lunch, fountain and dinner items available at any hour of the day that the operation is open (often 24 hours)

The Meal Plan and the Menu
- Patrons make two decisions- 1) What the sequence of food will be and 2) Which items will be selected- the meal plan must do the same- 1) Decide on the courses and their sequence and 2) Establish the specific foods in each course
- Food patterns have been changing- some are opting to eat less, more times per day, some are varying when the heaviest meal is consumed, so that varying from a strict meal plan offers welcome relief from monotony

Menu Organization
- Menus generally group foods in the order in which they are usually eaten
- Each menu category should offer choice- entrées are generally categorized according to the center of plate protein (Seafood, Meat, Poultry, Other etc.), methods of preparation should also offer variety; grilled, seared, fried etc. unless a particular method of preparation is the specialty
- Various tastes and textures afford further variety as do visual differences in color, cut and garnish

How Many Menu Items?
- The number can vary depending on operation- few are simple and helps control cost but may not offer sufficient variety. Many create complexities, cost control issues and, if too many, may perplex patrons
- Balance of tastes and selections; light-heavy, signature dishes and alternatives for vegetarians or non-meat preference, variety within like categories.
- 5-6 entrées are typical. Quality increasingly important if selection limited.

- Vegetables and starches with 5 and 3-5 respectively being usual choice range with variety in flavor, color texture etc.
- Side salad choices have decreased as entrée salads have increased with variety in ingredients and dressings. Salad bars seem to be declining due to costs and sanitation concerns.
- Fewer, more elaborate dessert choices are offered
- Breakfast items must be elaborate or unusual to justify cost- 30-40 items or more are usual

Menus for Various Meals and Occasions
- One menu with stated times for service of various items may be used or separate menus per meal period

  Breakfast Menus
  - À la carte and table d'hôte must be offered with light, "continental" breakfast to heavy meat and egg with starch meals
  - Other choices may include hot cakes, waffles, omelets, and other specialty egg dishes or meats.
  - Low calorie/fat/cholesterol items should also be available
  - Higher income items should be bracketed or highlighted to present to best advantage
  - Menu items may be numbered for easy ordering and are listed in the order generally consumed with large easy to read in the morning type
  - Bugger breakfasts may be for those in a hurry or with more leisure to linger when elaborate. Action stations where items are produced to order are sometimes featured.
  - Wedding breakfast may include champagne or gin for toasting
  - Hunt Breakfast is elaborate hearty buffet with meats
  - Chuck Wagon Breakfast originated from cowboy trail fare
  - Family style breakfast is brought in serving dishes for guests to serve selves
  - Group breakfasts must be planned carefully

  Brunch Menus
  - Combines items usually found on breakfast and lunch menus and provides for substantial meals.

  Luncheon Menus
  - May contain wide assortment of foods from complete meals to snacks
  - Moderately priced items call for higher volume and fast turnover to help cover costs though business and tourist trades may allow for higher ticket items and include alcohol
  - Lunch is generally the most flexible meal and may utilize specials
  - Group lunches may be fast paced or slow to allow for a presentation- regardless, the menu must fit the occasion and the group

Afternoon Menus/ Coffee and Tea House
- o Designed to catch the off peak trade, these menus offer snack and occasion food as well as caffeinated beverages, juices or "early bird" light dinner selections

Dinner Menus
- o Usually has more specialty items and must be carefully directed to patrons
- o Typically consists of appetizer, soup, salad, entrée with starch and vegetable and dessert with appropriate beverages though more casual family meals of sandwiches or pizza may be seen or menus which are geared toward the value driven guest
- o Service and décor are as essential as food in memorable dining- follow through must deliver all that the menu promises
- o Ethnic dishes or menus can deliver novelty and variety economically
- o Appetizers and desserts may be offered on their own menu- salads may come with the meal or prior to the entrée

Formal Dinner Menus
- o Lengthy coursed meals are seldom offered due to preference for less food and time constraints. Food is selected to give a flavor progression with heavy sweetness at the end
- o The most formal meal today typically does not have more than eight courses
- o Courses progress from; Cold Seafood, canapé or fruit, followed by soup, fish course, poultry course, roast meat course with starch and vegetable, salad, cheese and finally, dessert.
- o Wines and other alcoholic beverages must be carefully selected to enhance the dining experience not mask flavors. Sweet flavors are reserved for the end of the meal if at all, with as few as one or two wines considered to be sufficient.
- o The following characteristics assist in making a meal a memorable flavor event; Light and delicate appetizer and soup that refresh and introduce foods to follow, Bland fish, more pronounced, yet delicate and light poultry, followed by the roast, the peak of the meal. Salad should relieve and clean the palate; tangy cheese renews the palate and sets it for the sweet conclusion- dessert.
- o The most formal meals are served without bread or butter, salt or pepper. Less formal meals may have only 3-5 courses.

Evening Menus
- o Later night menus are geared toward those who have attended entertainment events or want to be entertained.
- o The offerings must be geared toward the market

Special Occasion Menus
- o Party and catering menus may follow a general theme or be specialized to the specific group or occasion
- o Suitability includes the ability to hold up to quick or delayed service and be feasible based on equipment and dishware

Party Menus
- o Vary according to the event and must have all costs carefully factored in
- o Work best with efficient systems of communication to all involved departments or areas
- o May involve a sales office with assorted menus developed for each meal period and occasion

Tea Menus
- o Low tea is simply tea with condiments or with dessert items, and/or fancy sandwiches
- o High tea is a light meal, sometimes served between lunch and a late dinner
- o Tea might be served as a reception
- o Coffee and tea house menus are gaining appeal and range from traditional to Bohemian

Reception Menus
- o Resemble teas but feature alcoholic beverages
- o Flying service refers to beverages and canapés passed on trays and is Russian in origin
- o Usual to estimate each guest will consume 2-8 pieces of food depending length of time, and variety of food offered
- o As canapés can be labor intensive, operations may purchase partially prepared portions or decorate only a portion. Guest desire for items that are not too messy/staining may also be a concern.

Buffet Menus
- o Original items as well as popular and comfortable items to suit the market
- o Number of items and presentation will vary depending on nature of buffet
- o Smorgasbord is a Swedish buffet with pickled herring, rye bread, and cheese and Russian buffets should have caviar, dark rye bread and butter with small glasses of vodka

Tapas and Tasting Menus
- o Spanish custom of "La Tapa" is eaten between main meals as small "snack" sized portions
- o Theories on origins vary. King Alfonso was ill and had to eat small portions with wine and declared something to eat must be served with wine at any inn. Another was that field workers needed snacks to keep them going until lunch.
- o Tapas include a variety of olives, dry nuts and cold cuts
- o Dim Sum is served in bamboo steamers or small plates so that a great variety could be sampled with traditional tea. Tea became popular during the Chung dynasty

Menus for Patrons Special Needs

Children's and Teenagers' Menus
- o Small children tend to have small appetites and limited items that appeal to them
- o Teens often have heartier appetites- their menus should be filling as well as appealing
- o Having a special Children's Menu is generally more successful than simply offering their selections on the regular menu. Items should be limited to favorites with simple wording and follow through on a theme. Activities or prizes that occupy children help in alleviating annoying behavior. Offering prizes that are part of a set or series encourages repeat business.
- o Menu items should be set for the geographic market. Regional favorites of one locale may not suit another.

Senior Citizens' Menus
- o Know the guest and their preferences. Seniors may fill off hours when offered discounts. Smaller portions of more easily chewed food and blander offerings have traditionally been offered.

Menus for Noncommercial and Semicommercial Establishments
- Captive markets must have nutritional considerations, preferences and variety kept in mind in institutional settings.

Institutional Menus
- o Operate from a budget to break even versus profit goal of commercial operations
- o Marketing and sales are generally not a focus as clientele are built in
- o Nutritional responsibility increases as options outside of institution decrease. Special diets may need to be considered.
- o Cyclical menus may add variety with seasonal or theme menus or by varying the meal plan
- o Most institutions use a three meal a day plan but some vary it with four or five meals, some of which are light. This may reduce labor costs and provide for patients who miss meals due to out of room testing and procedures
- o Changes in plans should be done carefully with input from patients/residents and adjustments if needed. Caloric intake should not increase
- o Hospitals are increasingly doing away with paper orders and using handheld point of sales systems.

Planning Institutional Menus
- Software programs that specialize in institutional menu planning have made this task much simpler. Calendar, cycle, shopping lists and printing recipes for forecasted quantities are all available options
- Variety if first established for entrées then vegetable, starches, salads and dressings.

- Balance, nutrition, costs and use of equipment and labor all must be checked

Health Facilities
- Menus needed include those for staff, nurses, doctors, visitors, catered events and of course, patient needs. Facilities include hospitals, convalescent centers, nursing and retirement homes
- General or house menu includes what food s will be served at a particular meal on a specific day. The dietician then selects, changes and substitutes items to meet dietary requirements
- Considerations include; diabetic diet, low-fiver, and low-fat.
- Retirement facilities resident rooms may contain kitchens where residents prepare some of their own meals, supplemented by a meal or meals prepared for a service in a common dining rooms. Less active residents may have rooms delivered to their rooms as in a hospital or hotel. Additional facilities such as a coffee shop and catering may be provided.

Business and Industrial Feeding Operations
- Various kind of service may include: executive dining room, staff dining, coffee shop and cafeteria.
- Meals may be at a subsidized cost or free, particularly in remote situations.
- Cycle menus have fairly long cycles as clientele does not change regularly. Printed menus with choices are most likely at the executive level. Cafeterias generally use menu boards.

College and University Food Services
- Many dorms have cafeterias with self bussing
- Meal plans may not allow credit for missed meals or may allow for the purchase of cards or fingerprint scanning for a prepaid amount. Per meal or d'Hote charges may apply
- Students often want substantial meals of popular food. Meeting nutritional needs and responding to desires are important.
- Student units may have a variety of food outlets including snack shops, quick service, seated service and banquets and catering.
- Faculty dining centers are also sometimes available. Lunch is the largest meal but dinner and bar service may be offered.
- Campus catering is often employed for a variety of functions

Elementary and Secondary School Food Services
- The majority of these schools are on the federal school lunch program and follow the mandated meal plan (a few are on the milk plan only)
- Plans must reflect the diversity of the student population to gain acceptance
- Teens use prepaid cards or point of sale purchases-some campuses are closed but regardless, foods must be popular with teens to be purchased

- Military and religious schools may be boarding schools with students eating all meals cafeteria, family style or table service
- A few offer separate faculty dining rooms with similar foods as students modified to adult tastes

Miscellaneous Institutional Food Services
- Orphanages, prisons, associations, religious and charitable groups may also provide food service, often on a very limited budget with government or institutional monies only

The Final Steps in Menu Planning
- Once all decisions have been made, pricing can begin
- Research into menu item popularity including seasonality must be continually done
- Menu analysis and financial considerations must also be ongoing

Questions for Review

1. A wedding breakfast could be differentiated from a regular breakfast menu by serving which of the following?
   a. Information regarding divorce
   b. Hot breads
   c. Bacon, sausage, and breakfast steaks
   d. Champagne or gin fizzes
2. Which of the following characterizes a hunt breakfast?
   a. Wild game
   b. Wine
   c. Lunch Menu items
   d. Buffet service
3. A high tea is one that features
   a. Imported tea
   b. Light meal items
   c. Little flowery teacups, saucers and adorable individual teapots
   d. Alcoholic beverages
4. The final step in developing any menu is to:
   a. Set menu item prices
   b. Analyze the success of menu items and the overall menu
   c. Choose equipment to be used in preparing menu items
   d. Evaluate the skill level of service employees
5. Which menu type offers food items separately at separate prices?
   a. Cycle
   b. Du jour
   c. Limited
   d. À la carte

6. The most formal modern meal usually does not exceed how many courses?
    a. People are in too much of a hurry for courses, they prefer bundled value meals
    b. 5
    c. 8
    d. 10
7. When planning institutional menus:
    a. Cost of menu items is not a high priority
    b. It is preferable to select foods which aren't prepared in bulk
    c. Presentation takes precedence over nutritional consideration
    d. Variety can be achieved by rotating menu items
8. Children's menus should:
    a. Have many selections of new and exciting items to help kids develop their palates
    b. Not increase the total amount of the check
    c. Provide appropriately kid sized portions of popular, familiar items
    d. Be written for parents viewings, not the child's
9. Several food items are grouped together at a single price in what type of menu?
    a. Limited
    b. Table d'hôte
    c. À la carte
    d. Cycle
10. A smorgasbord is a type of:
    a. Buffet
    b. Formal dinner menu
    c. Luncheon menu
    d. Breakfast service
11. In which of the following operations would a cycle menu be most appropriate?
    a. Professional Sports Stadium with season ticket holding fans
    b. Food Court located in a shopping mall drawing in tour buses and regular customers
    c. Cruise line offering week long Caribbean cruises
    d. National chain of quick service restaurants offering drive through window and delivery service
12. A restaurant that has elected to eliminate trans fats and utilize local organic produce from approved sources may be operating with which of the following common organizational objectives in mind?
    a. Employee welfare
    b. Diversification
    c. Ethical
    d. Management development
13. This type of menu is often associated with quick service restaurants or cafes and offers fewer choices than others:
    a. Limited menu
    b. Du jour
    c. California
    d. Cycle

14. Studies show that when people are confronted with a large number of menu choices they:
    a. Order more food and spend more money
    b. Tend to fall back on choices they have made before
    c. Want to return again and again to try everything
    d. Allow the server to make suggestions, are more satisfied and have higher check averages
15. Breakfast menus:
    a. Are best presented all on one page, thus smaller type is used
    b. Are at the time of day that guests are most prone to suggestive selling
    c. Should be in plastic page protectors as syrups and jams can create an untidy impression when smudged on paper menu pages
    d. Should be presented in large type as people are not yet fully awake

Discussion and Activity Questions

1. On the web or in your community, find an example of each of the following types of menus:
    a. Tapas, Dim Sum or Tasting
    b. Cycle
    c. California
    d. Children's
    e. Quick Service

2. Discuss each menu and rate it on the following dimensions:
    a. Appeal to intended market
    b. Variety, within what is reasonably expected for the category
    c. Health and nutritional concerns of intended market
    d. Cost concerns for the operation

3. What is the most memorable lunch can you recall purchasing or receiving in an elementary school setting? Explain your answer. Does this recollection have any validity in providing direction for today's institutional food service provider?

4. Of the illustrative menus used in Chapter Three, are there any that you feel could not work in your community? Explain why not. If all seemed to be feasible, which did you feel might have particular chance of success. Explain your reasoning.

5. Imagine you have been seated in a brand new restaurant and a flat, touch screen computer is brought to the table on which you are able to place your order. This interactive menu is capable of describing and showing each menu item in appealing detail, providing complete nutritional information, giving in depth answers on what, if any, modifications are available to each item and making suggestions in terms of complimentary beverages, sides, and accompaniments.
   a. Does this concept appeal to you?
   b. Do you think this concept has potential?
   c. In what kind of operation might this type of "invention" be utilized?
   d. Do you have any ideas for novel presentations of food choices to guests? Describe. Consider working any particularly inspiring notion into your menu project, if feasible.

Menu Development Activity

At this point, you should be firming up your menu concept(s) based on your anticipated market(s). Answer the following questions about each menu you are developing:

   a. What is the overall theme? How appropriate is it for your market? What would induce members of your market to find your theme compelling over all those other competitors out there?
   b. Is your menu covering a particular meal period or is it California style? What meal period is this menu intended to provide choices for?
   c. Will items be presented à la carte, table d'hôte, du jour or some combination or other configuration?
   d. What categories do you intend to present food items in and in what order? Will you select traditional titles for these categories or rename them to fit your theme?
   e. What items will be offered in each category? To what extent are ingredients cross-utilized?
   f. Is there a variety of textures, colors, flavors and preparation methods?
   g. Are there items for those who are health conscious? Are dietary restrictions or preferences for your market addressed with your menu choices?

Menu for Analysis:

The following menu has been reduced in size from the actual menu at Catch, Modern Seafood Cuisine located in Wilmington, NC. The author of the menu took this class using a previous edition of the text. Critique, using suitable modifications of criteria a-g..

Catch
Modern Seafood Cuisine, 215 Princess St., Historic Downtown Wilmington, NC
Starters

Charleston Crab Cake
Curry Potato Salad and Local Micro Greens **8.95**

Seared Yellowfin Tuna Summer Roll
Rice noodles, Salad greens, Pickled Ginger, Thai Basil + Vietnamese Dipping Sauce **8.5**

Firecracker Tiger Shrimp
Tossed in spicy cream sauce, fresh chives, on fresh lettuce nest **7.5**

Salads

Viet Chop Salad
Chopped Romaine, crispy calamari, cucumber, mint, sesame seeds and Red Curry Dressing **7.95**

Basic Green Salad
Mixed greens, cucumbers, grape tomato, carrot, croutons, & Feta cheese + lemon ginger dressing **6.25**

Smoked Shrimp Salad
Spring lettuce, curried almonds, dried cranberries and fried shallot + creamy roasted garlic dressing **8.95**

Saigon Salad
Fresh romaine, grape tomato, Thai basil, mandarin oranges, fried wontons, + Kaffir lime vinaigrette **6.5**
Add Salmon or Tuna 5

House Specialties

Grilled Tuna Sonoran
Steamed sticky rice, "Tex-Mex" black beans with mango salsa and Chipotle cream **12.95**

Broiled Miso Glazed Atlantic Salmon Filet
Veggie fried rice and sesame spinach **11.95**

Crab Fried Sticky Rice
California rice, Blue Crab, Chinese shallot, toasted garlic & bacon **7.95**

Torpedo Wraps
Flour tortilla filled with shredded romaine, grape tomato, spicy aioli and Viet sauce
With your choice of: **Shrimp 6   Flounder 7.5   Smoked Pork 6.5**

BLT
Crispy Applewood bacon, romaine lettuce, heirloom tomato and a zesty smoked pepper aioli on Flour tortilla
With crinkle cut French fries **7**

Tempura Philly Roll
Hickory smoked salmon, cream cheese and cucumber with sesame seaweed salad **9**

Fried Seafood
(Served with French fries, coleslaw and hushpuppies)
Choose from: Jumbo Shrimp **9.95,** Atlantic Flounder **7.95**
Choose 2- **12.95** Choose 3- **14.95**

For your health and ours, Catch restaurant is smoke free!
We support organic farmers, local fisheries and sustainable fishing practices
Seafood is a natural product; availability is limited due to "Mother Nature"

Chapter Four: Considerations and Limits in Menu Planning

Objectives
1. Describe the cost constraints in menu planning and explain what considerations relate to cost.
2. Describe labor constraints in menu planning and explain what considerations relate to labor.
3. Describe food purchasing constraints in menu planning and explain what considerations relate to availability.
4. Describe patron expectations and preferences and explain what considerations relate to variety, psychology, and health concerns in menu planning.
5. Explain how Truth in Menu Standards relate to menu planning and patron expectations.

Outline

Introduction
- Menu creator rarely has free reign to create a menu but must take into account a variety of constraints and constituencies. A menu must be in balance with all of its constraining factors.

Physical Factors
- The facilities must be capable of supporting production of menu items in expected quality and quantity. Facility assets must be utilized in an even manner.

    Equipment and Facilities Available
    - Menus must suit equipment capacity including the time to process it through
    - Storage capacity of refrigerated, frozen and dry storage must be considered for sanitation and cost factors
    - Creative problem solving with input from production personnel may provide assistance
    - Dining room service equipment must also be considered including the type of service and staffing

Labor Considerations
- Labor is one of the single biggest expenses in food service and must be thought of as a valuable asset
- Skills must be assessed using a *skills inventory* to neither under or overestimate the staffs abilities
- Training can overcome skill deficiencies. The menu must be challenging enough so that employees are not bored and have time efficiently used
- Food items must be available to be produced. Seasonality must be considered if a menu does not change often.
- Food item must be of expected quality
- Menu should maximize revenue and minimize costs. Non profit operations have cost and budgetary constraints that must be met.

Patron and Artistic Considerations

Guest Expectations
- Guests may not always know what they expect. Measures must be made through survey or focus groups.
- Wants and needs are different. The need for water when thirsty is a *physiological* need. The "need" to own a new television is a *perceived* need. Menus must try to satisfy both types of needs.
- Often needs and wants conflict. A menu planner cannot satisfy all of them.
- Offering healthy choices which do not sacrifice taste is an example of this. Guests know that they need to eat things that are good for them but do not want to experience deprivation.

Variety and Psychological Factors
- The theory of hedonism states that people try to maximize pleasure and minimize pain. As relates to food, this means that variety provides sensory input which is desirable
- Too much variety can cause dissatisfaction, particularly for those who are slow to experiment. Unusual preparation or presentation methods of common food can assist.
- *Satiety value* is the property of food that quiets the feeling of anxiety and restlessness created by hunger. Universities and prisons have found food to both relieve and create frustrations.
- There is comfort in eating at home so that environmental factors associated with dining away can create stress.
- Food may be associated with myth or have deep religious significance. These should be known and respected.

Appearance, Temperature, Texture and Consistency
- Visual factors such as color form and texture affect appeal. Color refers to food and dishes with colors associated with spoilage or unnaturalness being unappealing. Variety should be sought in the visual arena.
- Forms and shapes should be interesting and varied but not *busy*. Heights can be altered on the plate or through the use of dishes of varying heights
- Texture including shimmer of oil, jellies or liaisons or the assorted surfaces of food adds variety as does the creative use of garnishing.
- The *flavor* of food is distinguished primarily through aroma. The taste buds can discern sweet, bitter, sour and salt but our noses can detect far more. What is pleasing differs with individuals. Familiar flavors are most pleasing though many tastes can be acquired and savored.
- Our senses are sharpest around the ages of 20-25. As taste buds age, salt and sweet sensing nerves deteriorate first.
- Some cannot taste unless they also see the food. This is why presentation and display are important.
- *Texture* is the resistance food gives to the crushing action of the jaw. Complimentary textures enhance pleasure such as a crisp cracker with creamy soup.

- *Consistency* refers to the surface texture of food. Okra is slimy, for example. We do not cook vegetables for nearly so long as in the past.
- Temperature affects taste in terms of serving hot foods hot and cold foods cold. Cool cucumber salad, for example, enhances hot Indian curry.

Time and Seasonal Considerations
- Time of day, year and proximity to holidays can affect choices. Geographic locale affects availability of products and range of selections seasonally available.

Rating Food Preferences
- Preferences are influenced by food habits acquired over time. What people say they prefer is not always what they select.

The Taste Panel
  o Most use their own personnel to gauge appeal of various dishes potential for inclusion on the menu. Score sheets may be used.
  o Some have better taste perception than others. Smoking, alcohol and age affect the ability to judge taste.
  o Taste panels are best conducted at 11:30 or 4:30 when hunger is likely. Less formal sampling might be done with guests.

Patron Expectations
- The purpose of a menu is to communicate what is offered and, in commercial operations, the price. Patron expectations and what is served should coincide. Keep the language simple to avoid confusion. Descriptions should clarify without being overdone
- Exceptions exist but they are not the rule. Mike Hurst and Rich Melman have succeeded with menu items with names such as *Don't Ask, Just Order it*
- Terminology used must be accurate
- Items are generally named in the menu with a sub description, further explained by the server

Truth-in-Menu Standards
  - These laws, when in force, apply to menu accuracy and are for the purpose of avoiding misrepresentation which benefits the operator at the expense of the consumer
  - Purer Food, Drug and Cosmetic Act of 1938 forbid pictures or language descriptions that misrepresent the product.
  - The federal government has developed *standards of identity* that define what a food must be if a specific name is used and exist for all foods except very common ones such as sugar
  - On January 1, 2006 the USFDA came out with mandatory labels which will list the amount of heart disease linked trans fats and the presence of eight major potential allergens.

Health Concerns
- Another version of the food pyramid was released in April of 2005. It emphasizes the need to couple healthy eating with exercise. The chart can be customized on the www.MyPyrimid.gov website. The purpose of the pyramid is to help avoid the known risks of being overweight that increases the incidence of diabetes, heart disease and cancer.

Menu Pricing
- Menus must often be priced to meet customers expectation of value the guest and commercial profit or institutional cost coverage of the operation

Summary
- The perfect menu weighs and balances the constraining factors.

Questions for Review

1. A skills inventory is used to provide the operation with:
   a. List of skills needed for new hires
   b. List of skills current employees need and must be trained to perform
   c. An assessment of skills employees currently possess
   d. A job description
2. The senses are sharpest around the age of:
   a. 15-20
   b. 20-25
   c. 35-40
   d. 50-55
3. The following is true about creating a new menu:
   a. It is exciting to have a free hand at inventing a dining experience
   b. Costs are the primary concern above all others
   c. It is a balancing act, juggling the constraints of various constituencies
   d. Software is available that has taken the guesswork out every sort of menu planning
4. Commercial and institutional food service equipment
   a. Can produce in quantity but still has limits on capacity. The utilization of various sorts of equipment must be balanced and maximized.
   b. Can produce in mass quantities almost without limit. That is why the cost is so high for these products.
   c. Is best conserved by rotating the use of various pieces of equipment through daily specials and frequent menu adjustments.
   d. Have made amazing advancements whereby the food is almost prepared without human intervention. The initial investment is made up for by the labor cost savings throughout the lifetime of the product.

5. Efficient use of staff, challenging work and productivity:
    a. Are some of the reasons for high turnover in our industry. A more relaxed atmosphere would improve the workplace climate and solve many problems.
    b. Creates motivation and results in cost control.
    c. Are the pipedreams of the very inexperienced, idealistic managers that are sure to leave our industry for more soothing careers in banking.
    d. Are not nearly as important as the more compelling issues of energy const control and waste of paper supply products.
6. Healthy choices for guests:
    a. Are wildly popular. Our industry has been demonized for causing obesity.
    b. Can be popular if the guests do not feel deprived. They know they *need* to eat healthy, they just don't always *want* to.
    c. Are a passing fad. Before we know it, they'll be back to triple bacon cheeseburgers and chili cheese fries with a double thick malted.
    d. Must not be legislated by government. Trans fats are not so bad. Those too weak willed to pass the dessert display will die happy.
7. College and University students often show distinct ties between frustration and:
    a. Their studies
    b. Their significant others
    c. Their parents and grandparents who may encourage them to pursue careers for which they have no real interest.
    d. Food.
8. Many times institutional food is:
    a. Of much higher quality than perceived by those missing the home atmosphere
    b. Simply dreadful
    c. About the same quality as one would find in the household of any average family
    d. The motivation behind why some persons are in the institution in the first place.
9. Food can be best experienced:
    a. Blindfolded with the nose held, in order to truly focus on the taste and texture sensations
    b. When the senses are not confused by hunger, whereby anything tastes good.
    c. By senior citizens with the maturity to savor and appreciate a fine meal
    d. By seeing and smelling as well as tasting and particularly when one is hungry.
10. The theory of human behavior that says that people try to maximize pleasure and minimize pain is:
    a. Hedonism
    b. Bacchanalism
    c. The Peter Principle
    d. The Pyramid maxim

Discussion and Activity Questions
1. Go online and find a menu with appealing, healthy choices in the following categories:
   a. Quick Service
   b. Family or Casual Dining
   c. Bars and Taverns
   d. Fine Dining
   e. Delivery

2. Why is it difficult to be both healthy and appealing?

3. Examine the Healthy Options for Eating Out Guide in your textbook. Which of the tips may be unpopular with a restaurant operator? Explain. What might the operator do to meet the needs of the guest for whom this guide is aimed? How do the tips assist the operator?

4. What responsibility does our industry have when it comes to serve appealing, healthy food? Do some industry segments have more responsibility than others? Explain.

5. Write three menu descriptions for the following products- one that is misleading, a second that is accurate but unappealing, and a third that is both accurate and appealing:
   a. A grilled pork chop with peppers and egg noodles
   b. A piece of pie purchased from a local purveyor of fine dessert products
   c. A low fat frozen yogurt smoothie with numerous add-ins including chocolate chips, strawberry syrup, coconut shavings, and a ripe banana.

Menu Development Activity

Take the menu(s) that you have been working on and evaluate them in terms of production feasibility by your classmates, using the equipment and facilities available (or described by your instructor as available.)

Examine your menu for healthy appealing offerings. Are there options available to meet a variety of needs?

Work on descriptions of your menu items following the guidelines for truth in menu and simplicity set forth in the text.

Menu for Analysis
Analyze the following menus for:
   1. Are there healthy and appealing offerings to meet a variety of dietary preferences within the perceived market? (Who do you perceive the markets to be?)
   2. Are the menu descriptions appealing? Do they appear to meet the Truth in Menu standards?
   3. How would these menus play with the Healthy Options for Eating Out Guidelines in your text?

Jo Jo's Funky Monkey Circus

Monkey Bites

Crunchy Monkey
Raisins, Rolled Oats, Carob Bits and Dried Bananas

Monkey Meals
Meals served with choice of cinnamon apple sauce or carrot and raisin slaw

Melty Monkey
Mild Cheddar and Muenster Cheeses melted over a whole wheat English Muffin

Noodle-o Monkey
Slurpy spaghetti with tomato sauce and choice of beef or vegetarian meatballs

Oink and Cluckin Rollups
Chicken breast strips wrapped in ham and roasted

Thirsty Monkey

Nana Smoothee
Deelicious Bananas blended with strawberries and frozen yogurt

Jungle Juice
Your choice of Grape, Cherry or Mixed Fruit Juice Box

Milk
Chocolate or Regular Carton

Yummy Tummy Monkey

Nana Split
Banana and a scoop of frozen yogurt with pineapple, butterscotch and chocolate topping

Banana Pudding
Your favorite flavor with an oatmeal cookie

Lil Podners Chuckwagon Roundup

Appetizers

Onion Rigs

Fried Onion Rings with Hot Mustard Dipping Sauce

Petite Pate

Child Sized Portion of our award winning duck liver pate

Entrées

Entrées served with fried potatoes and steamed turnips

Pan Fried Prawns

Bite sized shrimp are deep fried and served with wasabe dipping sauce

Crispy Flounder

Filet of white fish is fried and served with garlic mayonnaisse

Escargot

Garlic Butter Snails in Puff Pastry or over pasta

Drinks

Caffeinated Sugary Soda

Available in brown or clear flavors with syrup from our bar. Small sized served in a martini glass (just like dad) or in a frosted beer mug (just like big sister)

Dessert

Cheese Platter

An assortment of aged cheeses served with seasonal fruit

Grand Marnier Souffle

Our signature dessert served in a child sized portion. Allow 45 minutes for our kitchen to produce this delicacy

Ask your waiter for a pen to draw on your menu

Chapter Five: Cost Controls in Menu Planning

Objectives

1. Perform the following basic cost calculations: food cost, portion cost, recipe costing and labor costing
2. Describe what is meant by "Prime Costs" and explain how these cost factors affect menu planning.
3. Relate various control techniques to their associated costs and describe how these methods function.
4. Compare and contrast cost control factors unique to commercial and nonprofit institutions

Outline

Introduction
- Costs are probably the most challenging limitation placed on menu planners. Food and Labor are the primary costs but other costs add up as well. Operational costs must be known to stay within budget.

Obtaining Operational Costs
- All costs must be known to be controlled. Reports are used to analyze data but simply having information does not solve problems.
- In *Purchasing*, the right kind of food, cost, quantity, quality and supplier must be found.
- In *Receiving*, it is essential to determine that what was ordered is what was obtained.
- *Storage* controls involve procedures so that nothing is lost to spoilage, contamination or theft.
- *Issuing* must be exact
- *Preparation* and *Production* are controlled through standardized recipes and procedures.
- *Computerization* has made these processes more efficient and allowed data to be made available easily for analysis

Calculating Food Cost

Total (Overall) Cost of Goods Sold
- Beginning inventory + purchases – ending inventory + total food cost
- Food cost divided by total sales = food cost percentage
- Food cost can be broken down into categories such as dairy, fish and meats to help pinpoint problems
- *Precosting* is an estimate of future costs and helps with planning
- *Averaging* is another way of Precosting. It is used with limited menus separated into similar or like items in each group. Sales mix must be known to do this.
- *Budgeting* is an estimate of future total food cost

- *Spindle method* is the simplest way of compiling total food costs using all purchase invoices. This method does not consider inventory so is only an estimate. Computerization is eclipsing this old fashioned method.

Individual or Group Item Food Cost
- The cost of food items can be obtained from the purchase price per portion or item, calculations of the number of portions, recipe costing and yield tests

Calculating Portion Cost
  o Divide total cost by number of portions obtained

Calculating Portions Obtained
  o Portion costs = cost of the unit divided by number of portions less losses through shrinkage, and increases due to additions of other ingredients

Recipe Costing
  o Divide recipe yield into total cost of cost of recipe plus determined percentage 2-10% for incidentals such as seasonings, fat, garnish or crackers.
  o Computer software greatly simplify these calculations and can perform a variety of functions
  o Portion costs are sometimes combined to include the entire plate costs including all accompaniments
  o Salad bar costs are derived by dividing the number of patrons into total cost of the salad bar giving an average cost per person

Conducting a Yield Test
  o Made to see how much edible food is obtained from raw, unprocessed items plus cost per portion
  o Simple yield test involves testing a recipe to see how many portions are produced and adding up ingredient costs
  o Complex yield test involves trimming meat with credit for usable byproducts and debit for labor cost so comparison with portioned product may be made
  o Edible Portion (EP) costs from As Purchased (AP) includes loss due to shrinkage in cooking (and carving)
  o Produce AP and EP costs vary due to a variety of factors such as product age (tender young may lose more weight than coarser textured old)

Obtaining Labor Costs
- This is a major cost factor for commercial and noncommercial operations.
- Elaborate preparations can result in high costs due to time and skill
- Good hiring, training and making employees feel valued

45

Monitoring Labor Costs
- o Simple preparation methods or the purchase of convenience foods can save labor if quality meets expectations
- o Service requirements also add to labor cost including tableware cleaning and storing
- o Wages, salaries, payroll taxes, employee meals and other costs and benefits make up labor cost. Commercial operations calculate dollar cost and then state it as a percentage of sales. Labor dollar cost divided by Sales dollar cost = Labor cost percentage
- o Labor cost per meal, per item, per patron or average cost per employee may be used
- o *Direct labor cost* is the cost involved in directly producing, serving or otherwise handling a menu item
- o Direct labor cost plus food cost for a particular item = *Prime cost* or total cost
- o Costs will vary based on the type of operation. Clubs may have a much higher cost than take out units, for example

Controlling Costs
o Costs are generally budgeted. If a menu consistently fails to meet budget, it may not be feasible for use.

Institutional Cost Control
- o Percentages are generally not used as often as in commercial operations; instead a budgeted allowance is used. Institutions may be subsidized with costs other than food and labor minimized.
- o Costs may be based on per meal, per bed or per person per day.

Commercial Cost Control
- o Cost allowances are usually based on a percentage of sales with 35% being typical for food cost and 30% for labor, though these may vary
- o Precosting through the use of computers has simplified the task of determining sales mix and reasonably predicting costs

Controlling Food Costs
o Many factors can cause high food costs

Forecasting
- o Forecasting is a prediction of how many of each menu item will sell to determine if profits will be adequate
- o Various types of food service will experience rhythms of business activity corresponding to times of the month, holidays, days of the week etc.
- o The firmer the basis for the estimate, the better the quality of the forecast
- o Weather can affect a probability forecast

- o Promotions can swell volume
- o Good forecasting can indicate how much food to produce, poor forecasting can result in overproduction, waste and loss or underproduction with decreased guest satisfaction

Portion Control
- o Management must establish and enforce portion sizes to control food costs
- o Standard portion sizes established as customary may be used or found in standard recipes used for food service
- o Standard dishes, portioning tools or volume can be used to assure that portions meet standards
- o Watching portions and preparation are standards to train
- o Pan weights should be carefully worked out
- o Foods should be marked or scored to indicate portions when possible
- o Items can lose weight and volume in cooking so that shrinkage must be factored into procedures

Selecting Items to Meet Cost Need
- o High priced items may be substituted, have other associated costs reduced or combined with other lower cost foods in a table d'hôte menu
- o Purveyors may assist in finding items to meet needs including price restrictions
- o Trade offs may be made by balancing higher cost items with lower ones if popularity is balanced

Controlling Labor Costs
- o Control methods include; allocation of labor cost per unit, per # meals per day, per hour, per covers, or per dollar sales. Commercial operations allocate according to a percentage of sales
- o Labor budget is made based on the above
- o Scheduling is crucial in control as is selection, training and motivation of employees
- o Time clocks or computers can be used to record work time and more

Allocating Costs
- o Some operations allocate labor based on dollar sales
- o Another method is to allocate positions based on dollar sales amounts or check covers
- o Staffing formulas have been developed from which requirements can be established
- o Quantity of front and back of house labor varies based on the type of operation
- o Labor budgets might allocate given hours for various departments per period

- o Formulas can give broad estimates of requirements
- o Number of employees will be greater than number of positions to allow for days off, part time etc.

Scheduling Labor
- o This task is often poorly done resulting in ineffective cost control
- o *Work schedules* show when workers are to be on the job
  - o Each employee and time at work for specific period
- o A second type shows days off and vacations
- o *Production task schedule* for workers on the job is a third type that can be combined with the first
  - o Shows work to be done by who
  - o Amount to produce using what recipes, portion sizes
  - o Meal or completion time required
  - o Comments and slack time assignments
- o *Bar Schedule Graph* helps determine if schedules match predicted business volume
- o Days off are important to employees and should be known at least a week in advance so they can make plans and the operation can meet needs
- o Vacation time can also be planned for times when labor requirements are less with planning
- o Teenagers hours and times are controlled in part by federal regulations. The type of work may also be restricted.
- o Overtime is also an important to avoid through schedule control due to the high cost of pay
- o Forecast and actual figures should be compared to evaluate
- o Computer programs can be used to budget labor

Improving Productivity
- o Controls in labor waste are directed toward; hiring, training, labor saving devices and foods requiring less, improving layouts, forecasting and scheduling labor
- o Poor selection results in inadequate performance particularly when training is not provided
- o *Job Specification* indicates qualifications, skills and experience of employees for specific positions and should be used in interviews
- o *Job Description* indicates tasks of each position
- o Training and evaluation helps employees know how they are to perform and how to improve
- o *Work Simplification or Human Engineering* is the field of knowledge about how to improve jobs which may increase productivity and morale

Other Controls
- o Fixed costs should be examined to see if they are, in fact, *Programmed Costs* which could be controlled or changed. Energy, utilities and supplies are some of these
- o Employee meals may save money through reduced pilferage and improved sanitation. They are labor cost if the price is below prime and not charged to employees
- o Being willing to sell out prevents waste through over prepping or serving an inferior product
- o Employee empowerment through corporations allowing manager buy in/ownership creates personal responsibility and reduces turnover and its associated costs
- o Part time employees lend flexibility, enthusiasm and reduce overtime

Budgeting Costs
- o Budgets should be based on realistic expectations not hopes. They should be flexible to changing conditions but not so subject to them that they no longer serve a control function

Ration Allowance Budgeting
- o Institutions use allowances per person. They may be based on the USDA with levels including low, moderate and liberal.
- o Budget should not be based on past performance alone as present conditions may not be equivalent
- o *Zero Based Budgeting* is the method of not using any past figures so that the current period may be well researched and planned

Questions for Review

1. Why must operational costs be known to the menu planner?
   a. It is their duty and responsibility to have as much information as possible
   b. In order to stay within a price budget or cost allowance
   c. So that the planner knows what level of funding is available for décor and renovation
   d. So the planner knows what kind of bonus and fees might be reasonably charged for services rendered
2. If an operation has a beginning inventory of $89, 345.00, ending inventory $86, 493.00 with purchases totaling $117,464.00, what is the total food cost for the period?
   a. $120,316.00
   b. $175,838.00
   c. $29,345.00
   d. $293,202.00
3. If the same operation had total sales of $356,321.00, what would the food cost percentage be?
   a. 82.29%
   b. 8.24%
   c. 49.35%
   d. 33.77%

49

4. Based on this information, it may be fair to conclude that management;
    a. Should be given a gigantic bonus for performance beyond what is standard in industry
    b. Be immediately fired and cautioned never to seek employment managing a food service operation again, ever
    c. May be running a commercial operation within the industry standard
    d. Is clearly in charge of a hospital, military or educational operation that is heavily subsidized and operating reasonably efficiently
5. Estimating costs is commonly done using;
    a. The spindle method
    b. Precosting, utilizing daily sales, averaging and other information
    c. Consultants
    d. The Comparative Formula software program
6. If a recipe for curry carrot soup costs $34.67 for the ingredients and yields 66 portions, what is the per portion cost?
    a. $2,288.22
    b. $31.33
    c. $1.90
    d. About 53 cents
7. Labor cost is;
    a. A relatively minor consideration in menu planning
    b. A necessary evil
    c. A major factor in planning for both commercial and institutional operations
    d. Relatively minor for institutions as they often have access to reduced or cost free labor, while still a big concern for commercial facilities.
8. Convenience foods tend to;
    a. Reduce labor costs and raise food costs
    b. Increase labor costs and food costs
    c. Reduce labor costs and food costs
    d. Increase labor costs and decrease food costs
9. Labor costs associated with service;
    a. Tend to be low, as servers are paid below minimum wage, due to tips
    b. High as there are so many more front of the house than back of the house staff to pay
    c. Are dependent on the type of operation and service provided
    d. Difficult to pinpoint in menu planning as chef planners largely calculate food costs and leave service to dining room management to control
10. Types of schedules typically used include;
    a. Production work, days off and vacations and work schedule
    b. Bar schedule graph
    c. Station assignments and slack time tasks
    d. Deductions and tax brackets of positions employed

Discussion and Activity Questions

1. Go online and search for commercial/institutional convenience food products in the following categories:
   a. Portion cut Rib Eye steak individually packaged in sealed plastic
   b. Soup
   c. Lasagna or pizza
   d. Appetizers
   e. Desserts

   What criteria would you use to decide whether to make or buy these products?
   What advantages and drawbacks did you find with each product?

2. What are the prevailing wages for kitchen staff in your geographic area? Servers? Is there a shortage or surplus of qualified workers?

3. Contact two entrepreneurial food service operations, one institutional and two commercial chains. Try to approach a range of types of operations. Ask the following questions:
   a. What sort of selection process is used to determine which applicants are hired?
   b. What sort of training is provided to new hires and for staff already past their 90 day initial period of employment?
   c. What sorts of incentives or benefits are offered to induce employees to continue working there?
   d. Does the operation use standardized recipes?
   e. Is everything on the menu made from scratch?

4. How would each of the following impact a manager's ability to keep costs within budget? Explain strategies that could be used to control costs in each incidence.
   a. An increase in minimum wage of $1.00 per hour for all hourly and tipped employees
   b. An increase in the price of lettuce, coffee, and shrimp
   c. A number of new operations have opened offering appealing employee benefits. As a result, a number of hourly employees are leaving for "greener" pastures.

Menu Development Activity

Take the menu(s) that you have been working on all along. Provide recipes for each menu item, cost them out and determine what each recipe yields. Forecast how many of each menu item will be ordered and explode recipes, costs and yields accordingly.

Hint. Recipes for commercial food service may be found online with exploding options available. If any of the convenience products located in the earlier activity seems suitable, consider substitution for made from scratch.

Menu for Analysis

Analyze the menu for any challenges that might exist in the area of cost control. Discuss each challenge in detail including strategies you might use to overcome these challenges.

Dining de Ultra Fine

Appetizers

Freshly Shucked Oysters on the Half Shell
Served with fresh lemon and horseradish

Steamed Clams
Fresh clams steamed in white wine with garlic and fresh herbs

Salads

Caesar Salad
Prepared tableside with classic dressing

Fresh Tropical Fruit
Papaya, Pineapple and Mango are freshly cut and served at the peak of ripeness

Entrées

Lobster Crepes
Prepared tableside from the lobster(s) you select from our tank

Steak Diane
Flambéed tableside with fresh wild mushrooms

Shellfish Soufflé
Freshly shucked or steamed shellfish baked in a soufflé with fine herbs and served with grilled baby organic vegetables

Quail
Spit roasted quail selected by you from our own back barn pen and served with haricot verts and fingerling potatoes

Desserts

Cherries Jubilee
Fresh organic cherries flambéed tableside and served over our own hand cranked vanilla bean ice cream

Baked Alaska
A mound of own hand cranked vanilla bean ice cream covered with meringue, doused in liqueur and brought flaming to your table

Chapter Six: Menu Pricing

Objectives
1. Characterize the most common pricing techniques used in the foodservice industry.
2. Identify historical perspectives on pricing and explain why non-cost pricing methods are no longer practical
3. Discuss the factors of pricing that influence product selection, including value perception and economic influences.
4. Explain how pricing psychology and market research play a role in menu pricing.
5. Characterize the most common pricing techniques based on costs used in the foodservice industry, including food cost percentage, pricing factor/multiplier, prime cost, and actual costs.
6. Characterize other pricing techniques in the foodservice industry, including gross profit, one-price, marginal analysis, cost-plus-profit, minimum charge, and pricing based on sales potential.
7. Identify software's role in assisting with food pricing.

Outline

Introduction
- Prices must cover costs
  - Noncommercial operations may have a safety net with government or charitable contributions to make up shortfalls
  - Commercial operations profit averages 3-7% of sales, so the margin of error is narrow
- Market is a major factor in pricing
  - *Perception of value* will vary from guest to guest
  - Different operations use different *markups*-the difference between the cost of a product and its selling price

Non-cost Pricing: Historical Perspectives
- These methods not based on costs were sometimes successful but are often risky

  Pricing Based on Tradition
  - This refers not only to a specific price but also pricing structure and market
  - Customers may resist when prices have been established for a long time
  - Market leaders may set prices and others of the same sort follow but this is impractical if costs are not factored in.

  Competitive Pricing
  - Common method whereby prices are based on what competition charges
  - This does little to *differentiate* and may not be cost based
  - A study of competitors prices, how they may be achieving them and how they affect your operation may be useful
    - *Value bundling* done by Quick Service is an example of a strategy that has been used in response to customer demand

What the Market Will Bear
- o Some operations design a product and then test market prices for acceptance
- o Developing a wanted product and pricing according to the value *in the minds of patrons* will develop a strong market, say experts
- o What the Market Will Bear pricing is popular and legitimate and will be enhanced when attempts are made to show guests true value
- o Differentiated product is a strong method of enhanced guest value perception

Value Perception
- What patrons think of desirability compared with price
- High Price may be associated with high quality and low price with lower quality
- *Differentiation* is the development of special, unique characteristics in a product and allows the seller better control of the market and pricing
- Food service uses special food and beverage creations, location, atmosphere, décor and service to acquire customer loyalty

Pricing Psychology
- Studies have been done regarding how patrons react to menu prices
- Planners try to avoid whole numbers
- "5" is the most common ending digit followed by "9" for Quick Service menus
- Fine dining sometimes spells out whole numbers using reverse psychology ("Twenty-Six", "Nine" etc.) to represent dollar amounts
- Four digits appear to be far more expensive than three digits
- Ranges of prices should not be skipped without careful consideration nor should too wide a range be used or customers will select lower priced items
- Price thresholds exist beyond which it is difficult to breach

Market Research
- Research should point to what kind of market exists and what consumers will pay
- A *base price* and *top price* can be defined with the menu working within this range

Economic Influences
- Laws of economics and commerce benefit and harm business
- When supply is limited, prices tend to rise and when plentiful, tend to drop
- When demand is high, prices rise and when low, drop
- As it is difficult to restrict supply, restaurants must seek to create high demand
- Dropping prices can increase demand and may make up loss with increased volume
- Buying seasonally can allow prices to drop when items are in demand
- *Market price* is used for products such as seafood where the cost is volatile
- Advertising and special promotions can increase demand
- When supply and demand increase and decreases in relation to prices the market is called *elastic*, if it does not, it is *inelastic*
- Products that are purchased regardless of how high the price is, or are not regardless of how low represent an *inflexible* market; if demand rises and falls with falls and rises in price the market is *flexible*

- A market where neither the price, nor supply and demand are changing is called *steady*
- Menus must reflect and take advantage of the influences of economic laws

Pricing Methods

Pricing Based on Costs
- Listing costs of food for each menu item and then marking up to obtain a selling price is one of the most common methods of pricing
- Disadvantage is that it assumes other costs remain the same with every menu item which is not necessarily so
- Using simply food costs also may not consider hidden food costs such as spoilage

Derived Food Cost Percentage
- Divide dollar cost of food by the desired percentage

Pricing Factor or Multiplier
- Food cost percentage can be converted into a *pricing factor* or *multiplier* by dividing the desired food cost percentage into 100 percent. This factor is then used to multiply food cost to get a selling price.

Variable Cost Pricing
- Food cost markups are assigned and labor is divided into (H)igh, (M)edium and (L)ow with different percentages assigned to these, thus food and variable labor costs are weighed

Prime Cost Pricing
- Prime cost is raw food cost plus direct labor, obtained by timing work

Combined Food and Labor Costs
- For this method, the labor cost must use a dollar value based on the cost for that specific item

All or Actual Food Cost Pricing
- Also referred to as *pay-yourself-first* is used in operations that keep detailed and accurate cost records divided into food, labor and operating cost units
- A desired profit percentage is also established
- This method can only be used by an operation with a good cost accounting system

Gross Profit Pricing
- Gross profit dollar figure is taken from the P&L statement divided by number of guests served to get an average dollar gross profit per guest
- Useful because in many operations, costs of serving each patron much the same after food costs and evens out prices in a group

The One-Price Method
- A doughnut shop is an example or a nightclub with cover charge or where selling food is not the primary purpose such as a casino

Marginal Analysis Pricing
- Objective method in which the maximum profit point is calculated with the selling price establishing the maximum used, perhaps by Quick Service

Cost-Plus-Profit Pricing
- Standard profit per customer is desired so a set amount is added to total costs

Minimum Charge Pricing
- As in Cost-plus, the rationale is that every customer costs a certain amount to serve and having a minimum charge will cover these costs
- Menus will state that there is a check minimum. Clubs often require members to spend a specific minimum which they are charged for even if expenditures fell below in order to remain in operation.

Cover Charge
- Set price added to bill regardless of what menu items are purchased. Also used by nightclubs with entertainment or dancing.

Pricing Based on Sales Potential
- Factors in addition to food and labor are considered
- Items are divided into (HC) high cost, (LC) low cost, (HR) high risk or (LR) low risk and (HV) high volume or (LV) low volume.
- LC, HV, LR are considered favorable while HR, HC an LV are considered unfavorable with pluses (+) assigned to favorable factors and (-) minuses assigned to unfavorable.
- Highest markup is any item having three or two minuses, lower one minus with the lowest for three pluses.

Pricing Aids
- Tables and computer printouts and software programs assist in performing accurate calculations

Evaluating Pricing Methods
- Few operations use only one method of pricing- most use a combination to best meet their needs and their patrons
- Pricing is not done and then over but must be constantly evaluated to study customer reactions, cost data collection and follow-up

Pricing for Nonprofit Operations
- All previous information is applicable to nonprofit except that most do not need to allow for profit in their prices but to cover costs and perhaps allow for a bit of a safety factor
- Accurate cost information is needed to break-even
- Estimates are made with allowances given to cover costs to budgeted amounts

Pricing Employee Meals
- Costs for food service where employees pay or companies subsidizes them must be calculated so they can be covered.
- Employees meals are often considered a benefit and therefore an operating expense
- Value may be based on actual cost, experience, standard or arbitrary charge
- When employees eat same meals as patrons, costs may be arrived at by estimating the number of employee meals, calculating individual meal cost and considering food as 50% of the cost of the meal. Finally, the number of meals is multiplied by the food cost per meal to arrive at a nominal menu price.
- Deductions should only be made for meals that are actually consumed/taken

Changing Menu Prices
- Customers may resent changes, particularly for items with good acceptance
- When food prices increase rapidly, customers recognize that the operation must cover costs but when food prices are stable, it seems less justified and therefore less acceptable
- One method is to remove an item and return it with the new price so it is less noticeable or making the change when the volume is down
- Prices are often changed when new menus are printed
- Using menu clip-ons eliminates the need to print new menus just to incorporate new prices- prices should never be crossed out
- Gradually working prices up may be more acceptable or using an announcement or table tent to explain
- If changes are frequent, only a few should be made at a time, more if changes are infrequent
- *Seasonable* or *market price* should be used for items that need to be changed frequently
- Price changes should be done within the range expected by patrons

Pricing Pitfalls
- Pricing should not be based entirely on just one cost like food
- Pricing should not ignore the forces of supply and demand
- Value perception should have greater emphasis in pricing
- More attention needs to be given to market information in establishing prices

Questions for Review

1. Break-even is the goal for:
   a. The first year of any food service operation
   b. Non-profits
   c. Commercial operations with multiple outlets
   d. Any year following a loss of revenue

2. *Bundling* refers to:
   a. Packaging for hot food to be delivered, generally in the pizza home delivery food markets
   b. Serving bread with the meal for no extra charge, generally in Fine Dining
   c. Offering drink, side and sandwich, for example for a price below what the individual items would cost, generally in Quick Service.
   d. Charging the same markup for food items in the same categories

3. Non-cost pricing methods:
   a. Are not based on cost and have been historically used in pricing menus. They are considered risky today.
   b. Are not based on cost and have been historically used in pricing menus. They are considered the standards by which pricing decisions are made.
   c. Are based on an assortment of factors and are considered a sound way to make pricing decisions in today's business environment.
   d. Are the newest techniques in the industry for pricing menu items profitably

4. Basing prices on what the competition charges:
   a. Is a great way to build a competitive advantage
   b. Is a way to tap into what the market will bear and therefore a known and proven formula
   c. Is wise to pay attention to but unwise to completely make pricing decisions as your operational costs may not be identical
   d. Assists in differentiating your product from that of others

5. A guest who happily pays $6.00 for fried onions cut in a clever fashion and served with an economical yet tasty dip, which costs the operation about $2.25 in combined food and labor costs is demonstrating the principal:
   a. You can charge almost anything for bar food
   b. A differentiated product with a high perceived value gives the seller better control of the market and pricing
   c. Hungry people will abandon nutritional and budgetary goals for good old fashioned fried food
   d. It is possible to run a 37.5% Prime Cost

6. Ignoring the fact that one price is the lowest, psychologically speaking, which price would be the most appealing to a Quick Service customer?
   a. Four
   b. $3.99
   c. $3.95
   d. $4.25
7. If a restaurant offered liver and onions for $1.00, and still, it remained an unpopular item, the market for the product would be:
   a. Elastic
   b. Inelastic
   c. Inflexible
   d. Flexible
8. The disadvantage of using food cost as the sole factor in determining a markup is:
   a. It assumes all other costs associated with preparing menu items remain the same with every menu item.
   b. Food prices rise and fall according to season and availability requiring menu prices to change frequently and driving up printing costs
   c. The primary cost factor is labor, which is not factored in
   d. It is likely the market may pay more and thus profit potential is robbed
9. Variable cost pricing is used:
   a. When food prices are volatile
   b. To test the market for price acceptability
   c. For special groups contracting for large functions
   d. For à la carte menu items and to calculate prices for high, medium and low labor costs items
10. A frozen dessert unit that charges the same price for a one scoop cone of ice cream, regardless of flavor ordered and the slight variation in actual costs of flavors is using the:
    a. One-Price Method
    b. Marginal Analysis Pricing
    c. Prime Cost Method
    d. Simple Cost Method.

Discussion and Activity Questions

1. Obtain menus from the following types of food service operations:
   a. Your educational institution dorm or student cafeteria
   b. A Quick Service restaurant
   c. A fine dining establishment
   d. An all-you-can-eat buffet
   e. A catering company

   Now, answer the following questions about each menu:
   a. What is unique about the pricing for this operation?
   b. What factors do you suppose influenced the pricing decisions?
   c. Do you believe the prices represent value to the target market?
   d. Do you have any suggestions to make regarding pricing for this type of establishment?

2. Look in your local newspaper and your mail for foodservice coupons or promotions. What effect do these have on pricing strategy? What are the benefits? What are the risks?

3. Review the last three months of your own experiences dining out. Where did you go? What were the determining factors in your decision to patronize a particular restaurant? How much did price impact your dining decisions? If you were to win the lottery, would you continue to eat at the same establishments? Why or why not?

Menu Development Activity

Take the menu(s) you have been working on and develop prices for each item on the menu. Keep in mind your objectives. Is the intention to break-even, or make a profit? If the menu is produced at an academic institution as a part of your class, are some costs subsidized such as energy, mortgage and labor? How does this affect your pricing?

Menu for Analysis

Analyze the following menu for pricing strengths and weaknesses.
What market do you feel is targeted with this menu?
What strengths and weaknesses do you find outside the realm of pricing?
What changes would you make to the menu and why?
If you wouldn't make any changes to the menu, explain why you made this decision.

Country Cookin' Buffet
All you can eat- Taste the South in your Mouth!

Cole Slaw
Potato Salad
Tossed Greens Garden Salad
Ambrosia Fruit Salad
Cottage Cheese

Yeast Rolls
Cinnamon Buns
Biscuits

Mashed Potatoes with Red Eye Gravy
Macaroni and Cheese
Sweet Potato Casserole
Green Beans with Field Peas
Cauliflower with Cheese Sauce
Fried Okra

Pulled Pork Barbeque Low Country Style
Steak cooked to order
Fried Catfish
Fried Chicken

Make your own Sundae
Coconut Cake
Brownies

All you can eat with doggy bag	$9.99 per adult
All you can eat with no doggy bag	$7.99 per adult
All you can eat to-go	$.59 per ounce
All you can eat kids	$4.99
Beverages- all you can drink	$2.99

Sodas, Sweet tea, Milk or Coffee

Ask us about catering your next event!

Chapter Seven: Menu Mechanics

Objectives

1. Identify the basic requirements to make a menu an effective communication and merchandising medium.
2. Describe the services offered by design firms and the considerations associated with determining a good fit.
3. Discuss aspects of using type: typeface, type size, line length, spacing between lines and letters, blank space, weight and type style.
4. Indicate how to give menu items prominence by using displays in columns, boxes or clip-ons.
5. Indicate how to best use color in menus.
6. Discuss paper use, construction of covers and other physical factors.
7. Indicate how menus are commonly printed, how to work with professional menu printers and the various methods of self-printing.

Outline

Introduction
- *Communicating* and *selling* are the main functions of a successful menu

Menu Presentation
- How a menu is presented does much to convey the type of operation and food service
- Most menus are printed on paper and given to patrons to look at, though not all
- A cafeteria board may list items and prices for patrons to view while waiting in line
- A Quick Service operation might have backlit menu boards with pictures behind the counter or at the drive-through
- Handwritten menus give a homey and personal touch
- There are a dazzling variety of presentation methods to accomplish goals
- Manner of presentation should best meet the needs of the operation
- Some will have a number of menus: breakfast, brunch, lunch, etc. or bar, coffee shop, and main dining room
- Most common menu is presented on firm paper with the front used for design, logo or motif, inside à la carte offerings and back may have additional offerings or information about the operation and locale. These items are permanent
- Menu items that change may be printed on lighter paper and attached to the more rigid menus
- À la carte and side dishes and their prices may be listed
- Specials can be attached as clip-ons or inserts. There should be space provided for this purpose so that no regular items are covered.
- Clip-ons or inserts are for the purpose of giving greater emphasis to items offered for variety but should not repeat items already listed. Color can focus attention.
- Specials are often listed on a board as guests arrive and then servers verbally describe them to guests at the table. When they number more than a couple, guests should have a printed version with prices.

- Menu Format
  - Readers should quickly understand what is offered
  - Simplicity helps avoid clutter
  - Foods should be on the menu in the order they are eaten except in a cafeteria where they might be on the order they appear on the line.
  - Locations should be included on menu boards where different counters are used for different foods.
  - Having representatives of the target market give feedback on the effectiveness of the menu in communicating may be helpful
- Production Menus
  - Never seen by customers, these are written for the back of the house workers to inform them what to produce
    - Sales terminology is not necessary, rather amount, recipe number, prep time, and other pertinent details

Menu Design
  - Design of a menu contributes greatly to legibility and patron reaction
  - Menus should have personalities and reflect the atmosphere and feel of the operation
  - Patrons should quickly grasp what is being offered at what price
  - Menu is an indication of what is to come; inviting the guest to a pleasing experience without promising beyond what can be delivered.

- Working with Designers
  - There are firms that specialize in establishing product brand
  - Simple logo designs to total brand identity and web design may be included ranging from $12,000-20,000
  - Complex e-commerce systems can be developed at a cost starting at $25,000 and up
  - Printable versions of the menu are often featured but need to be up-to-date
  - Companies with in-house designers and programmers may eliminate the high cost of needing to scrap a bug-ridden site
  - Finding a good fit is important to the success of the project

- Nontraditional Menus
  - Whatever is used, it must fit with the operation concept.
    - Lucite or Plexiglas for a modern effect
    - Natural materials may be keeping with the concept
    - Computerized screens may also be utilized where appropriate

- Using Type
  - Typeface
    - Fonts are the style of type used.
    - Serif is used often and deemed easiest to read. Letters are slightly curved
    - Sans Serif may look blocky

63

- Hundreds of faces are available but use of more than three on a single menu looks cluttered
- Print comes in plain, bold, italic and script- italic or script are most difficult to read but may create desirable special effects such as elegance
- Typeface should reflect the personality of the restaurant

- Type Size
  - Important to understanding and speed of reading- too large takes up space, too small is hard to read
  - Type size is measured in points which come 72 to an inch. 10-12 point type is used by most designers for listing menu items and 18 point for headings
  - Menus printed in a single size can be monotonous. Varying sizes for listings and descriptions is common
  - Headings may be in capital letters or bolder and larger type or different type if complementary
  - Sometimes emphasis items are listed in larger type or by boxing the item with a border and/or color contrast

- Spacing of Type
  - Space between letters in a word and between words. If too close or far, it is hard to read
    - Horizontal character width in a particular font might be condensed, regular or extended
    - Vertical spacing between lines of type is called leading- if none, the type is set solid
    - Thickness of leading is measured in points. For ease of reading, leading should be 2-4 points larger than the typeface being used

- Weight
  - Indicates the lightness or heaviness of print
  - Normal or medium is used on menus with bold or extra bold used for emphasis
  - Too light is hard to read, too much bold creates clutter and may clash with an elegant dining room. Light level should be considered also.

- Use of Uppercase and Lowercase Letters
  - Uppercase gives emphasis and can words out more clearly
  - Lowercase is easier to read
  - It is proper to capitalize first letters of proper names and main words in item titles. Proper nouns include menu item names.
  - Menu items are often put in large bold caps to stand out with lowercase type for descriptive material
  - Descriptions should inform as well as entice and keep with the menu theme and typeface compatibility.

- Special Effects Using Type
  - Script that matches the theme including ethnic restaurants can reflect a restaurant's cuisine

- Page Design
  - Page layout and design is an essential element of menu development. A good menu will lead guests to items operation wants to sell and a poorly designed menu will confuse them.
  - The amount of copy and use of space affects how quickly a menu can be read and understood.
  - About half of a menu page should be space
  - If more space is needed, more pages should be added rather than crowding
  - Avoid menus so large they are difficult to hold. Extra panels can add space or extra menus for desserts, alcohol and wine

- Line Width
  - The width of a line affects reading comprehension; 2/3 of those surveyed prefer double columned pages to single
  - 22 picas or just under four inches is the preferred line length
  - If a line is too long, the eye can jump, losing the place. One authority suggests putting prices right next to or directly under the item for that reason.

- Emphasizing Menu Items
  - Separating items by space and special type and then describing can help give emphasis
  - In addition to the margin space, one fourth to one third of the printed area should be blank
  - The first and last items in a column are seen first and best and therefore the place to put items one wishes to sell
  - Mixing up prices makes guests look through all items to see what there is and even price conscious guests may see something desirable on the basis of something other than prices
  - Patrons opening a two page menu focus on the middle right page and then upper right, left and down, across to lower right and up again. The areas readers look first are known as emphasis areas
  - Emphasis areas need not be preserved for specials as guests will look for these regardless

- Color
  - Can make an artistic contribution and affect legibility and speed of reading
  - Black print on white is read most easily though black on light colors such as cream is nearly equal to black on white.
  - Colored inks are also studied for readability
  - A large amount of color, while dramatic can be very costly
  - Considerable decorative effect can be obtained with only a small amount of color and design as with china

- Too much color and design can be distracting
- Clip-ons can be used to give a different design and color. This can also be achieved with ribbons, cords or tassels
- Variety in colors of papers including metallic is wide
- Colors can emphasize special events and holidays such as Chinese New Year done in black and red, St. Patrick's Day in green and Halloween in orange and black
- Colors should blend with décor or contrasted with good effect
- Menu designers are presumed to have a color sense that many laypersons may not and may be employed
- Menu publishing software may have complementary colors built in
- To reproduce a picture at least four pieces of film called separation must be made. Sophisticated machines now do separations by laser beam
- A four-color separation can cost more than $500 depending on size, paper and quantity ordered but the quality might be worth the expense
- When using photographs, the quality must be good, look good and strongly resemble the item prepared in the establishment
- Professional photographers or stock photos might be used
- Sketches and line drawings might also be employed for illustration. Decorative borders can add color and design.

- Paper
  - The Menu Cover
    - Paper should be chosen carefully. If disposable, such as a placemat it can be lightweight, if used regularly, coated grease resistant stock is important.
    - Heavy paper cover stock is used for most menu covers; stiff enough to be held without bending
    - Laminated covers last longer and they can be wiped clean.
    - Some operations use hard cover folders in which a menu is placed. They can be very decorative and expensive, padded, and covered in plastic or other materials. Foil inlays may be stamped in.
    - With these covers, the menu can be printed on lighter weight paper, perhaps self-printed

  - Characteristics of Paper
    - Interior paper can be lighter weight than covers
    - Paper can be given different finishes or specialty papers can be used.

  - Paper Textures
    - Textures can range from slight rises to a rough coarse surface.
    - Opacity (inability to see through it) may depend on the strength of the ink or the use of color. Most often, a maximum opacity is desirable.

Menu Shape and Form

- Can help create interest and sales appeal
- When using food shapes, they should be recognizable to the average diner
- The fold may also create an effect. Papers "foldability" should be investigated so cracks that affect many coated papers do not appear.
- Special shapes or folds require special dies- forms used to cutout shapes- will have to be made
- Dies and film or plates used for printing should become the property of the person paying for them. This must be spelled out in the original agreement

Printing the Menu

- Development of Typesetting and Printing
  - Chinese began the process of setting type in 700 AD. They carved blocks of wood and rubbed ink over the top surfaces.
  - Gutenberg invented movable type on the printing press in early 1400s Europe. This hand typing was long and laborious.
  - The linotype machine enabled a person to stroke a keyboard. This made the process faster and easier.

- The Modern Printing Process
  - Today, typesetting has become highly computerized. The type is reproduced either on transparent film or on photographic paper
  - The galley proof or reader proof is made, usually in long sheet to proofread and note corrections. These are made on the computer.
  - An artist will wither take the finish proof and make up the menu pages or they may be made up in the computer.
  - Page proofs with the illustrations and everything else in place must be checked. A photographic process is used.
  - Most printing today is produced by the offset method where the paper never touches the plate.
  - Silk screening is another method whereby a stencil is made either photographically or by hand and applied to a silk screen where ink is forced through onto paper. It is usually a hand process and expensive but to effect.
  - Hot foil stamping, embossing and die cutting are other methods of printing and enhancing a menu. Cost must be weighed against effect produced.
  - Permanent menus are printed in sizable lots usually. If small, prices may be left off a portion to be filled in at a later date. Most printers want a minimum order of 500. The need to change prices must be considered.

- Self Printing
    - Many operations today do their own printing at considerable savings while retaining complete control to change
    - A computer and good laser color printer are desirable
    - Care should be taken to create a unique menu that suits the operation identity
- Working with Professional Printers
    - Companies specialize in creating and printing menus
    - These companies have superior expertise in setting up menus to do the best merchandising job and avoid mistakes
    - Professional companies do special art and design work
    - Designers may work in conjunction with printers but it is always wise to seek references particularly when searching online

Questions for Review

1. No matter how well the menu is planned and priced, it must also be properly_____
    a. Shaped so that it pleases patrons young and old
    b. Presented so that it is understood quickly and leads to satisfactory sales
    c. Typeset so that fonts convey an elegant message of luxury and gourmet pleasures awaiting
    d. Folded so that unsightly creases do not detract from the menu items for sale to the hungry guest
2. Operations most likely to use permanent -paper menus handed out to guests include:
    a. Drive through windows
    b. Cafeterias
    c. Hospitals
    d. Fine dining establishments
3. A family restaurant would be likely to use menus that are:
    a. Presented in leather covers with tassels
    b. Laminated
    c. Handwritten
    d. Recited by the server from memory
4. Clip ons or inserts are used to highlight
    a. The logo, design or motif of the operation
    b. Specials or items that management wishes to sell
    c. Hours of operation and special information about the location or operation
    d. Items that might be overlooked due to a less advantageous location on the menu.
5. It is important for the reader of the menu to;
    a. Need to spend time reading and deciphering the menu so that they have a need to request assistance from the helpful server
    b. Be good and hungry so that the menu item descriptions entice them to buy
    c. Have a good understanding of food and wine terminology to assist them in their selection
    d. Be able to understand the menu and what is being offered at what price easily

6. The Menu Designer may offer services;
   a. As simple as logo design or as complex as creating brand identity
   b. That could be easily performed by most operators if only they had a simple software program
   c. And then flee to South America with the hefty fees, never to be heard from again
   d. In conjunction with a team of printers, programmers, recipe analysts and professional food stylists.

7. Type...
   a. Is really immaterial. A well written menu speaks for itself
   b. Is a complicated subject best left to the menu designers
   c. Is quite important in accomplishing the purpose of the menu as it affects legibility, reading speed and comprehension
   d. Can draw the diner to spend more than they had planned

8. Large menus...
   a. Can create an exciting experience for the guest, differentiating the operation
   b. May be bulky, awkward and difficult to handle if not done with care
   c. Assist in keeping the guests from socializing so much that they forget to order and tie up tables for excruciating amounts of time.
   d. Are difficult to store and irritating for height challenged hosts to distribute to guests gracefully

9. Patrons eyes travel first ...
   a. To the right center of a two page menu
   b. To the candle at the center of the table
   c. To the menu cover logo and graphics
   d. To attractive servers and dining room employees with kind eyes

10. The most important thing about photographs used in menus is;
    a. To obtain copyright waivers
    b. That good food photographs can be downloaded from the internet without all the fuss and bother of using a professional photographer
    c. That the photographs bring on a feeling of absolute desire to order and eat the food item portrayed
    d. That the photos be appetizing, quality representations of what it actually served

Discussion and Activity Questions

1. Go online and search for the following:
   a. Menu Designers
   b. Menu Software
   c. Professional Menu Printers

What did you find in each category?
What indication of quality, stability, and satisfaction did you perceive?
What kind of price range, if available, did you discover?
Did you draw any conclusions from this exercise?

2. Request copies of menus from your favorite restaurants. Examine them closely for adherence to the principles of this chapter. What strengths and weaknesses did you discover? Do any menus brought in by classmates stand out? Why?

3. If you were opening your own restaurant, what would you prefer to use for your own menus; Menu Designers, Self-Printing or Professional Printers? What are the advantages and disadvantages of each?

Menu Development Activity

Take the menu(s) you have been working on and design the physical layout, typeface, graphics etc. If available, use a menu software program such as SoftCafe MenuPro or a desktop publishing menu program. Simple word processing programs may also be used or an original pen and ink "master" copy. See how various sizes, fonts; weight and space effect the communication and selling ability of your menu(s).

Menus for Analysis

Examine the "Good Morning" menu as produced using self-printing menu software (SoftCafe MenuPro) in various styles (Caviar, Chinatown, Ginger Baker, Roaring Twenties and Strega Nona)

- Which menu do you believe best suits the perceived personality of the operation?
- Personality aside, which menu(s) make the most effective use of space?
- Critique each menu separately in terms of legibility and clarity, keeping in mind that each is identical in terms of content
- What color, paper finish, cover and treatments might you add to enhance each menu?

Good Morning Beverages

Freshly Brewed Coffee from just ground beans

Florida Orange Juice

Tomato Juice
With a splash of tabasco

Bottled Water

Lighter Appetite

Organic Vanilla Yogurt
Served with Fresh Bananas with Toasted Granola

NY Style Bagel with Philadelphia Style Cream
Cheese
Plain, Cinnamon or Multi Grain

Hickory Smoked Salmon Bagel
with pickled red onion and cucumber

Country Ham, Sausage or Applewood Bacon Biscuit

Breakfast Sandwich
Choice of meat with egg and cheese on country biscuit

Good Morning Breakfast

Unless specified, choice of Hash Browns, or Grits and Toast or Country Biscuit

Omelet

with cheese and your choice of breakfast meat

Platter

Eggs any style with your choice of breakfast meat

Shrimp and Grits

Smothered shrimp in low country gravy with stoneground grits

Fish and Grits

Fried Flounder, Grape Tomatoes and Stone Ground Grits

Vegetarian Breakfast Wrap

Scrambled Eggs, Navajo Black Beans, Mango Salsa and Jack Cheese

Sides

Stoneground Grits

Cheese Grits

Hashbrowns

Toast or Biscuit

GOOD MORNING

BEVERAGES

Freshly Brewed Coffee from just ground beans

Florida Orange Juice

Tomato Juice
With a splash of tabasco

Bottled Water

LIGHTER APPETITE

Organic Vanilla Yogurt
Served with Fresh Bananas with Toasted Granola

NY Style Bagel with Philadelphia Style Cream Cheese
Plain, Cinnamon or Multi Grain

Hickory Smoked Salmon Bagel
with pickled red onion and cucumber

Country Ham, Sausage or Applewood Bacon Biscuit

Breakfast Sandwich
Choice of meat with egg and cheese on country biscuit

BREAKFAST

Unless specified, choice of Hash Browns, or Grits and Toast or Country Biscuit

Omelet
with cheese and your choice of breakfast meat

Platter
Eggs any style with your choice of breakfast meat

Shrimp and Grits
Smothered shrimp in low country gravy with stoneground grits

Fish and Grits
Fried Flounder, Grape Tomatoes and Stone Ground Grits

Vegetarian Breakfast Wrap
Scrambled Eggs, Navajo Black Beans, Mango Salsa and Jack Cheese

SIDES

Stoneground Grits

Cheese Grits

Hashbrowns

Toast or Biscuit

GOOD MORNING

BEVERAGES

FRESHLY BREWED COFFEE FROM JUST GROUND BEANS

FLORIDA ORANGE JUICE

TOMATO JUICE
With a splash of tabasco

BOTTLED WATER

LIGHTER APPETITE

ORGANIC VANILLA YOGURT
Served with Fresh Bananas with Toasted Granola

NY STYLE BAGEL WITH PHILADELPHIA STYLE CREAM CHEESE
Plain, Cinnamon or Multi Grain

HICKORY SMOKED SALMON BAGEL
with pickled red onion and cucumber

COUNTRY HAM, SAUSAGE OR APPLEWOOD BACON BISCUIT

BREAKFAST SANDWICH
Choice of meat with egg and cheese on country biscuit

BREAKFAST

Unless specified, choice of Hash Browns, or Grits and Toast or Country Biscuit

OMELET
with cheese and your choice of breakfast meat

PLATTER
Eggs any style with your choice of breakfast meat

SHRIMP AND GRITS
Smothered shrimp in low country gravy with stoneground grits

FISH AND GRITS
Fried Flounder, Grape Tomatoes and Stone Ground Grits

VEGETARIAN BREAKFAST WRAP
Scrambled Eggs, Navajo Black Beans, Mango Salsa and Jack Cheese

SIDES

STONEGROUND GRITS

CHEESE GRITS

HASHBROWNS

TOAST OR BISCUIT

Good Morning
Beverages

Freshly Brewed Coffee from just ground beans

Florida Orange Juice

Tomato Juice
With a splash of tabasco

Bottled Water

Lighter Appetite

Organic Vanilla Yogurt
Served with Fresh Bananas with Toasted Granola

NY Style Bagel with Philadelphia Style Cream Cheese
Plain, Cinnamon or Multi Grain

Hickory Smoked Salmon Bagel
with pickled red onion and cucumber

Country Ham, Sausage or Applewood Bacon Biscuit

Breakfast Sandwich
Choice of meat with egg and cheese on country biscuit

Breakfast
Unless specified, choice of Hash Browns, or Grits and Toast or Country Biscuit

Omelet
with cheese and your choice of breakfast meat

Platter
Eggs any style with your choice of breakfast meat

Shrimp and Grits
Smothered shrimp in low country gravy with stoneground grits

Fish and Grits
Fried Flounder, Grape Tomatoes and Stone Ground Grits

Vegetarian Breakfast Wrap
Scrambled Eggs, Navajo Black Beans, Mango Salsa and Jack Cheese

Sides

Stoneground Grits

Cheese Grits

Hashbrowns

Toast or Biscuit

Good Morning
Beverages

Freshly Brewed Coffee from just ground beans

Florida Orange Juice

Tomato Juice
   With a splash of tabasco

Bottled Water

Lighter Appetite

Organic Vanilla Yogurt
   Served with Fresh Bananas with Toasted Granola

NY Style Bagel with Philadelphia Style Cream Cheese
   Plain, Cinnamon or Multi Grain

Hickory Smoked Salmon Bagel
   with pickled red onion and cucumber

Country Ham, Sausage or Applewood Bacon Biscuit

Breakfast Sandwich
   Choice of meat with egg and cheese on country biscuit

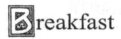Breakfast

Unless specified, choice of Hash Browns, or Grits and Toast or Country Biscuit

Omelet
   with cheese and your choice of breakfast meat

Platter
   Eggs any style with your choice of breakfast meat

Shrimp and Grits
   Smothered shrimp in low country gravy with stoneground grits

Fish and Grits
   Fried Flounder, Grape Tomatoes and Stone Ground Grits

Vegetarian Breakfast Wrap
   Scrambled Eggs, Navajo Black Beans, Mango Salsa and Jack Cheese

Sides

Stoneground Grits

Cheese Grits

Hashbrowns

Toast or Biscuit

Good Morning
Beverages

Freshly Brewed Coffee from just ground beans

Florida Orange Juice

Tomato Juice
With a splash of tabasco

Bottled Water

Lighter Appetite

Organic Vanilla Yogurt
Served with Fresh Bananas with Toasted Granola

NY Style Bagel with Philadelphia Style Cream Cheese
Plain, Cinnamon or Multi Grain

Hickory Smoked Salmon Bagel
with pickled red onion and cucumber

Country Ham, Sausage or Applewood Bacon Biscuit

Breakfast Sandwich
Choice of meat with egg and cheese on country biscuit

Breakfast
Unless specified, choice of Hash Browns, or Grits and Toast or Country Biscuit

Omelet
with cheese and your choice of breakfast meat

Platter
Eggs any style with your choice of breakfast meat

Shrimp and Grits
Smothered shrimp in low country gravy with stoneground grits

Fish and Grits
Fried Flounder, Grape Tomatoes and Stone Ground Grits

Vegetarian Breakfast Wrap
Scrambled Eggs, Navajo Black Beans, Mango Salsa and Jack Cheese

Sides

Stoneground Grits

Cheese Grits

Hashbrowns

Toast or Biscuit

Chapter Eight: Menu Analysis

Objectives
1. Identify the need for menu analysis and highlight criteria used prior to utilizing the menu and putting it into effect.
2. Explain the value of subjective evaluation in menu analysis and describe the level of knowledge desired in the performer.
3. Demonstrate how the popularity index or sales ratio is used in menu analysis
4. Demonstrate how to use menu factor analysis and describe the benefits
5. Identify the Hurst method of menu analysis and describe the benefits
6. Describe the value of the break-even method as an indicator of what a menu mast do to be profitable and demonstrate its use.
7. Identify other methods used by foodservice operations to analyze menus and describe their usage

Outline

Introduction
- After a menu is planned, priced and set into the form in which it is to be presented it should be analyzed to see if it:
    - o Meets the needs and desires of patrons in both the kinds of items offered and price
    - o Meets the needs of the operation by being feasible, profitable, and in line with the goals
- The menu cannot be considered final until its performance is measured. This is best done prior to printing unless self-printing.
- Even in institutional settings, items must be selling and should be changed if they are not

Common Methods of Menu Analysis
- The following seven methods shall be detailed in the chapter
    - o A count is kept of items sold per period
    - o A subjective evaluation is conducted
    - o A popularity index is developed
    - o A menu factor analysis is created in which performance is judged on the basis of; popularity, revenue, food cost and gross profit margin
    - o A break-even analysis determines at what point a menu will move from a loss to a profit
    - o The Hurst method of menu scoring determines how well a menu scores in sales, food costs, gross profit and other factors.
    - o Goal Value Analysis evaluates the effectiveness of menu items toward total sales and profits
    - o Each method can be tailored to the needs of the operation and any menu can be subjected to all methods as each has valuable information for management

Menu Counts
- A simple count of items sold is done, historically by having a clerk, cashier or other make a hand count tabulation
- A sheet was used which indicated how well items were selling and provided a summary of the sales mix
- Manual counts have become obsolete. Today electronic cash registers or POS systems record counts, food cost, gross profit for each item and total gross profit
- Computer does much of the basic work and can do a complete analysis using other methods.

Subjective Evaluation
- Simplest method uses independent expert review of effectiveness transmitting message and prediction of performance
- Value lies in the expertise of the analyzer. Personal views alone are not adequate.
- Forms can be used that are suitable to the particular type of operation being evaluated

The Popularity Index
- Menu items within a specific good group can compete against each other for selection with one killing the sale of others.
- Groups of food can compete against each other as well
- Relative popularity of separate items and different food groups should be known so their contribution to the sales mix can be estimated
- Potential of the menu as a revenue producing item and means of satisfying customer can also be estimated. A popularity index provides this information.
- Even noncommercial food services needs to study popularity and trends in patron selection should be known by both non and commercial establishments through a continuing record
- Not all menu items are expected to have high popularity. Some items are on the menu because management feels that they must be there.
- One method of compiling a popularity index is to:
  - Count of separate items is made within the group to be studied
  - All selections within the group are totaled
  - The percentage for each item of the total sold is calculated, the percentage obtained for each item indicates its popularity when competing with other items of its group
- A similar ratio can be calculated for groups of foods to analyze if they are competing negatively against each other.
- Calculations should be done for 30 days or longer to be informative
- Results should be carefully analyzed. Continued low ratio should be checked and reviewed for continued feasibility and price consciousness
- Figures can be misleading. Hidden factors should be examined; comments, day of week, or special events in or out of the operation
- Selling out of one item can force other selections so that a record of 86's should be maintained.

- o Weather or a sudden shift of popularity can also affect count as does menu placement etc.
- o Pricing and costing data history should be kept to evaluate selling prices
- o Magnitude of a popularity index depends on the number of items against an item competes. An item that is one in four is .25 (1 divided by 4), one in five, .20

The Popularity Factor
- It is difficult to compare popularity indexes of items when they come from groups that don't have the same number of items but are in competition with each other
- To remove this barrier, the actual popularity index (A) can be divided by the expected popularity (E) to get what is called a popularity factor. A popularity factor of one or more is popular, less than one, less so

Menu Factor Analysis
- A way of manipulating data to ascertain how well menu items are doing
- Factors indicate items popularity, creation of revenue or sales, influence on food cost, and contribution to gross profit. These are very valuable for management
- Actual indexes are derived from actual sales
- Expected indexes are derived from hypothetical projections given realistic expectations for items that are not expected to be equally popular- indexes could thus be equal or variable based on expectations. As in popularity factor, Actual is divided by Expected.
- 1.0 means the item is doing exactly as expected
- Menu factor analysis can be quite revealing and helpful for analyzing and indicating what could be done in case of problems
- Drawbacks include role of judgment in variable expectations, which could be flawed and lack of consideration of other factors.

The Hurst Method
- o Considers effects on sales of pricing, food cost, item popularity, gross profit contributions, and other factors
- o To use the Hurst method, management should decide on the period for which the scoring will be done. It should be a typical, not unusual period

Steps in Performing the Hurst Menu Analysis
1. Decide on the period and meal to be covered, number of items to be evaluated and fill in data
2. Select items that contribute to a major portion of revenue. The items are called the study group. Place these items in column 1.
3. Make a count on the number sold for each item and place results in column 2
4. Add entries in column 2 to get the total number of study group items sold
5. Add to the number of step 4 items and all other nonstudy menu items sold and place the total on the line for Total Items Sold and in space 11, Total Served
6. Record the selling price for each item in the study group in column 3

7. Multiply the number of each study group item sold (Column 2) by its selling price (column 3) to get the total dollar sales for each item. Record these in column 4.
8. Add column 4 to get the total dollar sales of the study group
9. Calculate the dollar food cost for each item in the study group. Record in column 5.
10. Get the total food cost of each sale by multiplying the number of each item sold by its food cost. Record in column 6.
11. Add column 6 to get the total dollar food cost for all items.
12. Obtain the meal average- or average selling price-of all items in the study group by dividing the total dollars in sales (the sum of column 4) by the total number of study group items sold (sum of column 2) Place the result in space 7
13. Calculate the gross profit and gross profit percentage. Subtract the total food costs for all study group items (sum of column 6) from the total dollar sales of study group items (sum of column 4) Record the difference in space 8. Then divide this difference by the total dollar sales (sum of column 4) to get the gross profit percentage. Record this figure in space 9.
14. Calculate the gross profit average by multiplying the check average (average selling price) in space 7 by the gross profit percentage in space 9. Record this in space 10.
15. Calculate the percentage of customers that select the study group items by dividing the sum of the study group items sold (total of column 2) by the total items sold as compiled in step 5 and recorded in space 11. Record the result in space 12.
16. Calculate the menu score by multiplying the gross profit average (space 10) by the percentage of patrons selecting the study group items (space 12) Record the result in space 13.
17. Make any comments in space 14
18. Sign and date the form

Evaluating the Menu Score
- One score by itself means little. A number of scores kept over a period of time must be compiled and analyzed to obtain desirable information on how well a menu is doing. Comparing scores with other operations is also not helpful.
- Often a low score is the result of too low a selling price or not being selected. Manager may try to reduce food cost or use merchandising to increase sales. Management must investigate carefully
- Computer can be used to simulate menu changes before actually putting them into effect as long as accurate data are fed into the computer

Goal Value Analysis
- It is largely a quantitative method using mathematical A x B x (C x D) = Goal Value

- A = 1 – Food cost percentage, B = Volume or number sold, C = Selling Price and D = 1 – (variable food cost percentage + food cost percentage) to arrive at the numerical target or score
- Individual values for each menu item is then calculated and compared with standard
- Goal Value Analysis can be very useful to foodservice operators however variable cost remains fixed in these calculations. In actual practice it does not. It can be made more effective if a high, medium or low variable cost is assigned.

The Break-Even Method
- Another way to analyze a marriage is to see whether it breaks even or covers all the costs of doing business. This can be done for profit and nonprofit food service. This can be good for commercial operations in the first year of business
- If costs are not covered, there is a loss, if more than covered, there is a profit. To know when a menu will break even, one can total all the costs of producing and serving the volume of meals and deduct this from the expected income.
- At times it is desirable to calculate break-even prior to putting the menu into effect
- The things that must be known to calculate break-even are: 1). Average check, 2) fixed costs in dollars and 3) the fixed cost percent
- Average check= total dollar sales divided by the number of guests served in a particular period
- Fixed costs are costs incurred regardless of whether a single sale is made. Rent, administrative expenses, license fees, depreciation, insurance, some labor costs, and employee benefits, heat, electrical power and advertising sometimes referred to as turn-key expenses
- Fixed cost percentage is the dollar value of fixed costs divided by sales
- BE=FC/FC %
- BE=FC/(AC x FC%) to calculate the number of customers needed to cover costs where AC= Average Check
- BE=FC+P (profit)/FC% to find the dollar sales needed to cover fixed costs and give a desired profit
- BE=(FC+P)/(AC x FC%) to find out how many customers would give this profit

Other Methods of Menu Analysis
- Computer printout of productivity reports are used to see how much of a certain item each server sold
- Gross dollar sale per seat, check of operating figures against a profit or a profit and loss statement
- Daily food cost report acts a check and often makes comparison with other periods

Questions for Review

1. How many methods of menu analysis are commonly used?
   a. There is really only one effective method; subjective.
   b. There are simply too many to count
   c. Seven
   d. Eleven
2. Today, menu counts are performed by:
   a. The cashier
   b. The chef
   c. The manager
   d. Electronic cash registers or Point of Sales systems
3. Which method of menu evaluation is simplest?
   a. Subjective
   b. Hurst
   c. Break-even
   d. Popularity index
4. In the popularity index method of menu analysis:
   a. Potential of the menu as revenue producing and customer satisfaction can be estimated
   b. Server popularity with guests is compared to sales
   c. Ranking of popular items is compiled and analyzed
   d. Preferences of guests who rank high in social popularity are analyzed
5. If an item is one of five and has equal popularity with the others, it would have an expected popularity of:
   a. 1.0
   b. .20
   c. 1/5
   d. 5
6. Generally a popularity factor greater than 1.0:
   a. Has a commendable food cost percentage
   b. Has not met sales goals
   c. Shows the item more than holds its own against others
   d. Has achieved break-even
7. In menu factor analysis, it is possible to:
   a. Have an equal and variable expected index
   b. Turn around a poor selling item by factoring merchandising techniques
   c. Hypothesize various successful menu scenarios
   d. Get cross eyed performing lengthy calculations
8. How many steps are there in the Hurst Method of menu analysis?
   a. 7
   b. 18
   c. 3
   d. 27

9. Goal value analysis is largely a:
    a. Subjective method of analyzing a menu
    b. Team effort involving management and crew setting sales goals
    c. Quantitative method of analyzing a menu
    d. Means of analyzing the soundness of an operations goals
10. If costs of doing business are not covered:
    a. There is a loss
    b. The operation is likely to be non-profit
    c. Someone is due a pink slip
    d. It may be the third year of the operation when typically sales do not keep up with expenses.

Discussion and Activity Questions

1. Which menu analysis method appeals to you most? Explain your reasoning.

2. Which menu analysis method appeals to you least? Explain your reasoning.

3. Interview three independent restaurant owners and ask them how they analyze the effectiveness of their menus. Report your findings to the class and compare notes.

Menu Development Activity

Using the menu analysis form in the appendix, analyze the menu(s) that you have developed, omitting criteria that you are unable to evaluate at this time. Do you have any new information to assist you in your quest to create a successful menu? What alterations might you make as a result?

Menu for Analysis

Regarding the Management by Menu Analysis Café below:

1. In each category, calculate expected and actual popularity factors.
2. What conclusions do you draw from these numbers? Explain your reasoning.
3. Calculate the contribution margins of each item.
4. What conclusions do you draw from these figures? Explain your reasoning.
5. Does the new information, coupled with data obtained from performing popularity factor analysis, change any of your earlier conclusions? Explain.
6. Numbers aside, give your subjective evaluation of the menu. What market might this menu appeal to? What changes might you make to improve it? Are ingredients cross-utilized efficiently? What other factors might have gone into the development of this menu?

MANAGEMENT BY MENU ANALYSIS CAFÉ

SOUP

Tofu Chili...$4.95
30 orders sold- 20% food cost

Cream of Roasted Garlic...$3.95
15 orders sold- 10% food cost

Gumbo with Seafood and Sausage.........................$4.95
45 orders sold- 45% food cost

SALADS

Chilled Thai Noodles with julienne vegetables...........$5.95
10 orders sold- 15% food cost

Fried Chicken Salad...$6.95
25 orders sold- 45% food cost

Anti pasta with sausage, marinated vegetables and cheese.....$7.95
15 orders sold- 30% food cost

Squid Salad- Marinated with Onions and Peppers.........$6.95
25 orders sold- 20% food cost

Nicoise Salad- Seared tuna, Green Beans, Hard Boiled..............$8.95
Eggs, Black Olives, Red Onions and Steamed Potato Cubes
25 orders sold- 30% food cost

PASTA

Seared Tuna with tiny new peas in 3-cheese sauce over............$8.95
Fettuccine
40 orders sold- 45% food cost

Calamari Diablo over Spaghetti...................................$7.95
15 orders sold- 20% food cost

Eggplant and vegetables Parmesan over fettuccine...................$7.95
25 orders sold- 20% food cost

Spaghetti marinara with Tofu "Meat"balls........................$8.95
35 orders sold- 25% food cost

Chapter Nine: The Beverage Menu

Objectives
1. Identify the basic requirements for planning an Alcoholic Beverage Menu.
2. Indicate how the Beverage Menu can be properly implemented through skilled merchandising and service.
3. Describe how to institute controls to ensure that the Beverage Menu satisfies guests and meets cost and/or profit goals.

Outline

Introduction
- No other phase of food service has changed so much in recent years
- Mothers Against Drunk Driving (MADD) has been a big factor in changing penalties and controls due to irresponsible consumption of alcohol and the heavy costs in lives
- Third party liability or Dram Shop laws were established so that those who profit from the sales of alcohol bear responsibility for selling it responsibly
- Insurance rates have risen accordingly
- Training programs are available from the NRAEF, AHLAEI and other sources
- A decline in drinking with more emphasis on quality than quantity has been a result
- Wine sales continue to be a growth market in the US. Flavored vodka sales have grown significantly and niche beers have found a market though beer sales have generally declined.
- The Beverage Menu has changed, including an emphasis on nonalcoholic beverages with a wider selection of wines and beers.
- Good Beverage Menus can help profits if;
    - Drink sales are promoted responsibly
    - Cost control procedures are established including inventory control
    - Menu choices that complement food
- Menu theories and principles hold here but the markup is often higher, alcohol is an accompaniment that must be merchandised and control procedures must be tighter. State regulations for the sale and service of alcohol must be observed.

Presenting the Beverage Menu
- Alcohol may be presented on the food menu when only a few choices are offered. Institutional menus may present this way. (Nursing homes or hospitals when doctors allow)
- Separate Beverage Menus may be used with one for mixed drinks and cocktails, one for the wine list and another with after dinner drinks. Afternoon, and Banquet are other possibilities
- No matter what method is used, the regular menu should announce the availability of alcoholic beverages
- Tabulation may be needed when choices are extensive but it is not necessary to have an exhaustive list of wine or beverages if choices are well thought out.
- As with food menus, Beverage menus should be analyzed periodically

- Expert help may simplify the task of menu development; purveyors, brewers and distributors can give assistance in terms of expertise, education and promotional materials allowed by state law.
- Beverage lists should be specifically designed for each operation, particularly ethnic establishments. Authenticity is important including glassware.
- Prices and descriptions should be clear and legible.

The Wine List
- 8-10 wines may be sufficient at a very casual restaurant
- 50-75 wines should satisfy even the most discriminating individuals. Guests anticipate a selection as exciting as their food choices.
- Premium wines by the glass are popular at most operations and may be offered in addition to bottles of wine
- Carafes may be used to enhance value perception. Flights are very small tastes equaling about a glass.
- Wine menus are difficult to design and require expertise. White and red dinner wines must be on the menu as well as dry wines from foreign or domestic sources depending on clientele.
- A distinction should be made between dinner, aperitif, sweet and effervescent wines.
- Dinner wines should be divided into dry reds, whites and less dry roses. Dry indicates lack of sweetness.
- Domestic generic wine indicates the kind of wine
- Varietal wines bear the name of the variety of grape- 75% of wine must come from that specific kind of grape by regulation
- Red wines are normally used at room temperature (60-65 F) with red meats, white dinner wines are served chilled (45-50 F) with seafood or poultry. These rules have faded to suggestions that guests order what they like.
- Sweet wines are proper after dinner and may be served in place of dessert.

      Merchandising Distinction and Variety
-     Wine stories or lore can increase appeal
-     Use of proper terminology is crucial

      Merchandising the Formal or Catering Wine List
-     Wine at a special event may be printed at the bottom of the menu or next to/below the course it is being served with
-     Wine may also be presented on a separate list on a folded menu with wine on the left and food on the right

The Spirit List
- Spirits are alcoholic beverages that are not wines, liqueurs, or beers that generally have a higher alcoholic content.
- The list is traditional with less variation than a wine list and may only list cocktails that management wishes to push with the understanding that the bar can create other choices.

Stocking Spirits
- The list should be balanced. Guests who order spirits generally know what they want. The purpose of the menu is to inform of prices.
- Call brands refer to the brand name of the spirits with the menu limited to those that move, with slow sellers eliminated. Local tastes and perceptions must be considered.

Proof
- Alcoholic content is measured in proof a term that originated with the British and refers to the combination of alcohol, water and gunpowder and the ability of the alcohol to burn, "proof" of the alcoholic content.
- In the US, proof is two times the percent of alcohol by volume or weight. 50% alcohol would then equal 100 proof.
- Wine is 7-14% alcohol, Beers 2.5-8%, and Spirits from 50-190 proof or 25-95% alcohol.

Beers and Ales
- A modest list would include 3-4 domestics and 3-4 imports. Ethnic restaurants often include beers from that country. If a beer list is offered, more beers will be featured.
- If several good draft beers are served, the need for a bigger list may not be necessary.
- Draft beers may be needed as well as popular light beers and local favorites. Knowing the clientele is crucial

After Dinner Drinks
- The list should be limited to what the guests require and no more. Clip ons, boxed listings or table tents can work well.

Nonalcoholic Drinks
- Offered for underage and non-drinking patrons can help increase the check average and offer an alternative.
- Nonalcoholic versions of popular cocktails, creative combinations of juices, freshly juiced fruits and vegetables, and nicely garnished sodas are all popular.
- Teas, coffees and chocolate drinks have also seen a resurgence in popularity and bottled or flavored waters sell well in still and sparkling versions

Wine Service
- A sommelier or wine steward discusses wines, makes suggestions and takes the wine orders. Servers and sommelier should know the wines and menu and be prepared to make alternative suggestions.
- Wine is enhanced with proper service. Red wine needs to breathe- older less than newer and white wines are generally served chilled, red at room temperature. Guests preferring different temperatures should be given what they desire.

- Presentation means wine label is shown to guests before the cork is pulled, foil removed and then whites put back in wine holder to continue chilling, reds on the table or into a basket. Cork is placed to the host's right.
- Host tastes the wine for acceptability. This task was traditionally performed first by the sommelier.
- Proper glassware should be used that allows the wine room to be swirled and breath. Shapes can be specific to the type of wine or all-purpose.
- Holding wine up to the light to check for visual qualities is part of the ritual.
- While the wine bottle is left for the guests, it is often set away from them so that servers can perform the serving of it

Beverage Service
- Poor selling and service can negate even a masterfully planned menu.
- Beverage service should be built around the emphasis of relaxation and pleasure. Friendly, helpful service by perceptive servers who have the ability to chat, make recommendations or leave the guest alone. Servers should have excellent product knowledge.
- Lounge service is different than dining area service. Lounge guests priority is drinking rather than the emphasis on eating in the dining room. Keen observation including the judgment not to over serve is important.
- Clean glassware, standardized recipes and attractive garnishes make a difference in sales and perception.
- Service should be prompt with techniques of suggestive selling responsibly employed.

Liquor Pricing
- Liquor markups are higher than food and averaging out prices is common methodology.
- A knowledge of cost is still desirable, as in all food service

Pricing Wines
- Wine prices should reflect menu prices
- Wine may be marked up 100% unless it prohibits sales. Beverages are more likely to be priced according to what the market will bear and/or based on competition
- Wine on a table will increase the chances that others who see it will order- therefore some operations price wine lower to move it. 50% over bottle cost is usually the bottom markup. Costs for wine include interest on inventory, storage, cooling, labor, glassware, and breakage which amount to 41%. This leaves 9% for profit and variable and fixed costs. If it is not enough, the price must go up.

Pricing Spirits
- Prices depend on costs; therefore exact quantities for drinks must be established with oversight that these limits are followed.
- Main ingredient cost is calculated with other costs added to it as in costing out any other recipe with the price then based on a 25% or other specified product cost
- Some operations estimate the added ingredient costs as 20% over the main ingredient cost

- 5% is allowed for loss due to miscalculation, over pour, evaporation and spillage for spirits and 7-10% for dispensing tap beer.
- Computerized or electronic dispensing of drinks may negate the need for these calculations.

Pricing Tap Products
- Draft products are usually priced based on unit cost. A 7% pouring loss is expected.
- Other pricing methods or charging only what the market will bear are means of pricing tap products

Beverage Control
- Control of alcoholic beverages is essential to profitability; over pouring, spillage, improper mixing, improper ringing up of sales, customers who refuse drinks or avoid paying, and employee pilferage are all ways profits can be lost.
- Control methods include; good purchasing, standardized recipes, proper production, adequate record keeping, and financial controls.

Purchasing Controls
- Menu dictates purchases. Alcoholic beverage purchases are simplified as most are bought by brand name or call brand.
- House or well items are good quality brands of spirits that are used for mixed drinks or others where no call brand is specified. These are often selected on the basis of local tastes and price.
- Price, preferences, quality and proof are factors in selection. If only a few order anything, it should not be stocked.
- Quantity purchased should be held to the lowest possible figure with safeguards against running out. Minimum and maximum level of stock should be set with safety factor to allow for delivery time.
- Discounts must be substantial to lure buyers to purchasing more than the maximum level. Storage, interest, risk of loss and other factors must be considered.
- Requisitions or purchase orders, including a daily receiving report are records that should be kept. Specifications should be consulted for receiving.

Receiving
- Items must be checked to see if they are the right item, size and quantity ordered. Errors should be noted on the receiving report and retained invoice. Major errors should be refused.
- Goods should be stored so that the oldest items are used first.

90

Issuing
- Issue of bar goods should only be done with a signed requisition with an empty bottle swapped for full. Some states specify how empties are to be handled.
- Full bottles that are sold to guests or transferred to room service or catering should be accounted for with a paper trail.
- Paper is increasingly being replaced with handheld computerized devices.
- Par stock should be the basis for ordering- or allowable inventory based on what management thinks is needed for a shift or day.
- Bar stock inventory is done in tenths or fourths of a bottle.
- Bin cards for checking off items removed from storage acts as a perpetual inventory but the computer is the preferred method of keeping track of inventory.
- Physical inventory should be taken by two people, sometimes with a handheld device. Some multiple outlet operations employ a single person to reconcile bar inventory.

Production Control
- Standard recipes for drinks should be as well tested and costed as food recipes
- When bartenders make drinks without measuring ingredients it is called free pour. While many guests and bartenders prefer this method, better consistency and cost control occurs when measures such as jiggers and shot glasses are used.
- Measuring caps can be put on bottles that do not allow more than a set amount to be poured. Counts of pours may also be performed by some caps.
- Push button pouring systems can be used to control drink amounts with detailed sales information recorded. As some guests do not care for this method, it may be employed at service bars out of sight only.

Service Control
- Point of sales systems have greatly simplified service control. Checks can be easily separated or split with a record of server and item sales
- When using paper checks, they should be numbered and issued to servers and bartenders
- Transfers from bar to dining room should be planned for
- Floaters are checks held by servers to be reused for the purpose of pocketing money owed the establishment. Use of a locked check box helps prevent this when the check is immediately placed there following payment.
- Whether payment is obtained with each order or at the end of service, the system used should serve the guests needs with most guests preferring to pay only once for beverages and food.

Reconciling Sales
- Methods that will allow a check on income received is called reconciling sales. A daily cost sheet presents what has happened but not necessarily what should be happening. Computerized point of sale systems have simplified the process.

  ### Sales Tally
  - o This method involves tallying sales from guest checks and checking the income against the amount of liquor used for that period

o   Point of sales systems have simplified this once cumbersome process.

Averaging Drinks
o   Average number of drinks sold from a bottle is used as a basis instead of the sales tally for income that should have been obtained.

Averaging Percentage
o   Average percentage involves a breakdown of the kinds of drinks from each bottle. The advantage is that they can be applied to any size bottle

Standard Deviation Method
o   A calculation is made to ascertain what income would be obtained if one kind of standard drink were sold at one price from a bottle. Very low or high variations should be examined with a normal variance expected.

Drink Differential Procedures
o   This method is sometimes used to allow for one type of liquor in a number of drinks sold at different prices. This can present a challenge, particularly when more than one type of liquor sold at different prices is used in a drink
o   Point of sales tallies help making an adjusted value per bottle
o   No system should cost more than the information it supplies is worth

Questions for Review
1. MADD stands for:
   a. Mothers Advocating Defensive Driving
   b. Mothers Against Drunk Driving
   c. Ministry of Alcohol, Dining, and Driving
   d. An organization of food service operators upset over escalating liability insurance rates
2. The strongest growth market in alcoholic beverage sales in the United States is expected to be:
   a. Beer
   b. Wine
   c. Spirits
   d. Coffee, Tea and Chocolate
3. The following statement about the sales of alcohol is True:
   a. The law on serving minors is strict and penalties can be severe
   b. It is rarely necessary to have a beverage menu as most customers know what they want
   c. Laws have eased up since the old dram shop law era
   d. An operation may set the hours for the sale of alcohol in their own establishment according to demand

4. The following is NOT a likely presentation for an Alcoholic Beverage Menu:
   a. Appearing on the back of the regular food menu
   b. Separate menu from the regular food menu
   c. Table tent with featured specialty drinks
   d. Lit signboard behind the counter
5. Drink menus are often necessary as guests want to know:
   a. Ingredients of their favorite cocktails
   b. The spellings of hard to pronounce wines
   c. Prices
   d. What specials are available such as two for ones, doubles for the price of singles, ladies drink free and other price discounts
6. What percentage of wine in the United States comes from California?
   a. 10%
   b. 25%
   c. 65%
   d. 95%
7. _____ has been referred to as a red wine masquerading as a white.
   a. Chardonnay
   b. Rose
   c. Riesling
   d. Burgundy
8. The average ingredient cost for liquor items is often desired to be:
   a. About the same as food
   b. 20-25%
   c. 30-40%
   d. 10% or less due to the added cost of liability insurance, security and specialty glassware.
9. If tequila is .75 per ounce, triple sec .50 and sour mix runs about 20% of the main ingredient total, what would a margarita cost that calls for 2 ounces of tequila, ½ ounce of triple sec and sour?
   a. $1.17
   b. $1.92
   c. $2.15
   d. $2.40
10. House or well items are beverages that are:
   a. Made on the premises
   b. Higher quality than call brands
   c. Stored in kegs
   d. Good quality brands of spirits that are used when no call brand is specified

Discussion and Activity Questions

1. What kind of restaurant appeals to you when you want food that you like and where you or your dinner companions can order alcoholic beverages that complement the food? If possible, bring in or print out a menu from an establishment of that fits this description.

2. What kind of lounge appeals to you when you want to relax with friends, drink an alcoholic beverage, and share some snacks? Is entertainment a factor? If so, what kind? If possible, bring in or print out a menu from an establishment of that fits this description.

3. Should the person who is consuming alcohol bear sole responsibility for the consequences of their actions? Why are foodservice establishments that serve alcohol being charged with liability when an intoxicated patron causes property or personal damages?

4. Develop a menu for a coffeehouse that includes food that goes well with your menu which emphasizes beverages.

Menu Development Activity

Take the menu(s) that you have been working on and develop a beverage menu suitable for the food menu. It may or may not include alcoholic beverages, depending on the market, meal period, and local regulations. Be prepared to explain your selections.

Menu for Analysis
Prepare a Beverage Menu for this restaurant's food menu:

Starters

Charleston Crab Cake
Curry Potato Salad and Local Micro Greens $8.95

Seared Yellowfin Tuna Summer Roll
Rice Noodles, salad greens, pickled ginger, Thai basil and Vietnamese dipping Sauce8.5

Firecracker Tiger Shrimp
Tossed in spicy cream sauce, fresh chives on fresh lettuce nest   7.5

Salads

Viet Chop Chop Salad
Chopped Romaine, Crispy Calamari, Cucumber, Mint, Sesame and Red Curry Dressing   7.95

Smoked Shrimp Salad
Spring Lettuce Leaves, Curried Almonds, Dried Cranberries and Fried Shallot with Creamy roasted garlic dressing8.95

Basic Green Salad
Mixed greens, cucumbers, grape tomato, carrot, Feta cheese + lemon ginger dressing   6.25

House Specialties

Grilled Tuna Sonoran
Steamed sticky rice, "Tex-Mex" black beans with mango salsa & chipotle cream   12.95

Broiled Miso Glazed Atlantic Salmon Filet
Veggie fried rice and sesame spinach   11.95

Torpedo Wraps
Flour tortillas filled with Smoked pulled Pork, shredded veggies, spicy aioli and Viet sauce   6.6

BLT
Crispy Applewood Bacon, Romaine lettuce, Heirloom tomato and zesty smoked pepper aioli 7

Chicken and Crab Fried Sticky Rice
Blue Crab, tender chicken, egg, shallot, toasted garlic and bacon   7.95

Chapter Ten: Producing the Menu

Objectives

1. Describe the relationship between planning the menu and the role of the purchaser.
2. Explain how purchasing needs are determined and how methods of meeting those needs are found.
3. Compare and contrast the various methods of bidding
4. Explain the relationship between specifications and quality standards.
5. Identify the purpose and benefits of purchasing controls and procedures used to facilitate them.
6. Explain the relationship between planning the menu and the role of production
7. Describe the purpose and benefits of production controls and procedures used to facilitate them.

Outline

Introduction
- The menu cannot stand alone but must be integrated with thorough purchasing and production procedures or the it will fail.
- The central document, the menu, is dependent on purchasing and production. The menu controls and directs but cannot act.
- Purchasing must correctly interpret what the menu calls for and procure the necessary materials in time to be produced.
- Production must prepare the food in a timely manner with due regard to food cost, quality, safety and quantity.

Securing the Product: Purchasing
- Purchasing is comprised of the following steps:
  1. Determine the need for an item along with the quality, quantity and other factors required to satisfy that need.
  2. Search for the item on the market
  3. Negotiate between the buyer and selling, ending in a transfer of ownership
  4. Receive and inspect, ending with the acceptance or rejection of the item
  5. Storage and issuing items
  6. Evaluate the purchasing tasks, as judged by the performance of the product, and the economy and efficiency with which results were achieved. * This important step is often overlooked.

The Role of the Purchaser
- Purchaser is to interpret what is needed to produce the items listed on the menu and procure them at prices within constraints established by the operation
- This is easier said than done because some buyers do not understand menu requirements, know how to satisfy, lack expertise to search out products or are only acting as order takers and givers.

- The job is complex. Issues include; fail to follow through with receiving and production procedures, being misled by bargain prices, and price paid by lost profits due to failure to meet quality and merchandising appeal factors.
- Markets can change quickly. Buyer must be aware and ready to move.
- Communication with kitchen ensures that the operation can respond proactively.
- Waste, legal and ethical woes can all plague an operation when the purchaser lacks integrity.

Determining Need
- Production lets the buyer know what products are needed by listing them on a requisition
- A reorder point may be based on addition of the established minimum par stock, a safety factor and an average usage amount from the time of ordering to the delivery.
- Standing orders occur when an established amount is replenished to the maximum. This can happen with dairy, coffee tea or bread, for example.

Searching the Market
- Call or quotation sheet is a copy of the list of established needs with the grade, quality, size and other requirements for each product are noted. Purveyors are called and asked to quote a price. Comparisons are made and a purveyor is chosen. The order is then placed.

  Bidding
  o Bidding is a formal method of buyer. Need is indicated to purveyors who quote prices. In the most formal type, specifications are submitted in writing followed by written bids based on requirements.
  o Samples may be submitted as a part of this process to help make decisions and compare delivered product with.
  o Deadlines also characterize formal bidding. A bid bond guarantees performance as expected. A failure to perform results in the bonding company forcing fulfillment or paying a forfeit. The cost of administering this type of program is worthwhile if the volume is very large (for example government, large enterprises and hospitals)

    Informal (Negotiated) Buying
    - The most informal bidding occurs when only a few items are needed and a minimum of three purveyors are contacted much as in call sheet buying.

    Cost Plus Buying
    - Purveyor buys items at cost with a specified markup added. Commodity items are most likely to be purchased this way.

    Blank Check Buying
    - Purveyor obtains items no matter the cost and buyer pays market price plus standard markup. Buyer must be reliable and trustworthy.

Writing Specifications
- o The heart and soul of purchasing is the delineation of what the buyer wants covering all characteristics and other factors to get the right product at the right price.
- o Buying should not take place until management determines exactly what is needed. Specifications usually include:
  1. Name of item
  2. Quantity needed
  3. Grade, brand or other quality information
  4. Packaging method including size
  5. Basis for price
  6. Miscellaneous factors required to get the right item
- o General specifications can simplify all others if delivery, billing, and bid acceptance requirements for all are delineated
- o Terms with precisely understood meanings must be known and used
- o Institutional Meat Purchase Specifications (IMPS) are known by meat purveyors and identify meat specifications by number
- o The North American Processors Association (NAMP) has published the Meat Buyers Guide to assist the operator with information
- o Known items are relatively easy to specify but items that are not well known are more difficult. Federal and state agencies have begun to develop specifications for convenience foods which can help.
- o IMPS and UPC are two code numbers often used in purchasing. UPC standardizes purchasing with a series of numbers with lines, these are used for inventory using Palm sized PDAs.

Quality Standards
- o Brand name is one way to assure consistent quality but are only as good as the manufacturers that make them
- o Grades are usually based on Federal standards and separate a product into different quality levels. Grades are established on known factors and do not generally change.
- o Trade grades may be used that have recognition only in specific markets
- o Federal grades are derived through consultation with industry and are issued as tentative grades. It is tested, revised if needed and then made official.
- o Levels provided to are: consumer, wholesaler and manufacturer. Consumer grades dominate the market with processors and producers following.
- o Federal grading of processed foods is based on scores for certain items with the total having a letter grade assigned to it, much like school grades.
- o Scoring methods have developed for different groups of food such as meats and produce.
- o Standard of identity is a statement from Federal Government that defines exactly what an item is by its characteristics. Terminology is also defined.

- These standards of identity become legal descriptions of the item allowing for easier negotiation between buyer and seller. Many are the basis for truth in menu
- Fill or weight standards also assist in buying and selling canned items, barrels, hampers, bushels and crates.
- Laws such as the Pure Food, Drug and Cosmetic Act, Federal Trade Commission Act, Agricultural Marketing Act and the Perishable Commodities Act protect buyers and sellers and give order, reliability and stability to the marketplace.

Purchasing Controls
- Good purchasing requires coordination, organization, and simplification of accounting, controlling and monitoring the flow of materials through an operation. Information that assists with accountability should be formalized by procedures.
- Paperwork can be handled by computer which simplifies and speeds the process, thus improving it.
- Various units may prepare requisitions which are put on Purchase Orders, call sheets or uploaded for the order to be placed. Each order carries its own number. Order lists and Pos should be properly authorized and executed.

Receiving
- This important step allows management to see that the goods delivered meet the terms in every respect described in specifications, purchase order and other purchase documents with proper cost and quality
- Discrepancies should be noted on the receiving sheet and invoice

Storage and Inventory
- Receiving personnel are responsible to move goods to storage as quickly as possible.
- Perishability means that food safety and quality may be compromised if food is not placed in proper storage immediately
- Prescribing delivery hours, when possible, assists with orderly processing and prompt storage
- First in, first out, or FIFO is a logical method of use arrangement for storage
- Minimum and maximum quantities should be established for each item. When supplies reach the minimum on the reorder point (ROP), a notice is given to the buyer to bring stock up to par or maximum
- Daily inventory may be used on foods on hand in kitchen or bar supplies
- Purveyors may supply forms or palm held PDAs with bar code software can streamline processes
- Perpetual inventory is derived solely from records. Present stock plus deliveries minus outgoing stock equals perpetual inventory. This is largely maintained by computer
- Physical inventory is derived by a physical, visual count
- Experience and knowledge are required to take a good inventory

- 1-2% difference in perpetual and physical inventories is considered normal but much variance should be investigated
- Meats are sometimes handled differently with a tag

Issuing
- Requisition forms are usually completed to indicate what items various departments want to withdraw from stores. Computer software might be used.
- Issues from the storeroom plus direct deliveries equal the amount of food used for time period data gathered; day, week or month

Value Analysis
- A review of the purchasing process to evaluate whether the best possible job has been done and determine whether it can be improved.
- Specific items purchased are analyzed, processed items are observed during production, and the supplier is appraised.
- Value= quality divided by price (Q/P=V)
- Yield must be known to truly obtain accurate portion costs
- Consolidation of orders by a group of operations may reduce costs and is called co-op or cooperative buying
- Guaranteeing payment within a specified time period may bring costs down
- Cost-plus buying or guaranteeing quantities are other methods as well as advance notification of needs to increase time to search for favorable buys
- Simplification reduces costs

Preparing the Food: Production
- Success of the menu depends on the skill of preparation of listings and successful production depends on thoughtful menu planning
- Workers must have skills and kitchen must possess equipment and layout conducive to successful production of the menu
- Good forecasts, recipe prep and portioning must be provided
- Flow of communication and controls must be established so that production sheet will serve demand
- Standardized recipes should be used and controls established to produce exactly the same item each time it is offered
- Costs must be accurately known
- Failure to inform results in an unprofitable menu due to haphazard work and disappointing product
- What is promised must be delivered
- Service personnel must check food and inform kitchen of discrepancies
- Most food items should be prepared as close to service time as possible unless advance prep enhances their flavor value
- Leftovers should be tracked as well as sell outs or "86's"
- Supporting units such as the butcher and pantry must be well organized to avoid disappointing guests by having menu items unavailable

Kitchen Staff
- Skill needed by the staff directly relates to the complexity of the menu
- Training employees in techniques as well as the underlying principles, assists in quality
- Personnel organization should be functional. There are traditional systems for various types of kitchens. Each should ensure that workers are supervised or assisted when they need help.

Control Procedures
- Labor can be controlled through the use of procedures such as production schedules, time clocks, payroll records, training records, and health records.
- Good forecasting, avoiding duplication of efforts and use of a computer can help tighten operations.
- Management should study the kitchen production system for efficiency. Procedures should be coordinated to bring the efforts of the team together so that items will be produced as planned at the right time.

Questions for Review

1. A purchaser who is knowledgeable but fails to see if the operation correctly receives and uses the product is referred to as:
   a. An order taker
   b. Market-oriented
   c. A buyer
   d. Unemployed
2. If the par stock for apples is three cases, the safety factor is two cases and two more are needed for usage between order and delivery, then the reorder point is:
   a. 12 cases
   b. 3 cases
   c. 4 cases
   d. 7 cases
3. When a list of needs is established, purveyors are called and prices are quoted for each, this most simple method of searching the market uses a:
   a. Call or quotation sheet
   b. Internet search engine
   c. Mystery shopping service
   d. Purveyor list search
4. The most informal bidding method is called:
   a. Blank check buying
   b. Cost Plus buying
   c. Negotiated buying
   d. Ante up

5. These have been all but replaced by the use of federal grades, but they still have recognition in specific markets such as fresh fruits and vegetables, dried fruits, eggs, and poultry:
   a. Universal Product Code
   b. Standard of Identity
   c. Specification
   d. Trade Grades
6. Quality can be defined in different ways. Two of the most common methods of identifying quality are:
   a. Brand name or grade
   b. IMPS number
   c. Choice or Prime
   d. Consumer ratings and taste tests
7. Much of the paperwork for purchasing procedures can be handled by:
   a. The Hospitality specialty accounting firm of Laventhal and Horwath
   b. Computer, simplifying, speeding and improving the process
   c. Methodical, filling out of forms available from vendors, online or through the corporate office
   d. The operation's bookkeeper
8. First in, first out or FIFO refers to:
   a. Seniority requirements of union workplaces
   b. Scheduling system for production workers
   c. Arrangement of item usage in which items are stored and used in the same order that they are received.
   d. The method of determining par level, safety and reorder points
9. A perpetual inventory is performed:
   a. Through a visual count
   b. Through a number of prompts such as meat tags, bar codes and time temperature indicator (TTI)
   c. Solely from records and usually maintained today by computer
   d. Daily (perpetually)
10. Standardized recipes ensure that:
   a. No chef feels that they are indispensable
   b. The exact same item is produced each time it is offered to a patron- consistent quality leading to satisfied guests
   c. There is no time wasted squabbling over the best way to prepare a menu item each time it is ordered
   d. The food item is recognizable to the serving staff who must pick up and deliver to guests

Discussion and Activity Questions

1. What might be appealing about the role of purchaser for a foodservice operation? What would be the biggest challenge to your skills and interests?

2. Contact the food purchasers for a large institutional establishment such as a hospital, a small entrepreneurship, and a local branch of a corporate chain and interview them regarding their purchasing duties, the challenges, rewards, and their thoughts on the profession.
3. Visit a local foodservice establishment that has agreed to allow you to view their receiving area, storeroom, and discuss their purchasing process.
4. Arrange to visit a local vendor or purveyor or have one speak to the class.

Menu Development Activity

Take the menu(s) you have been working on and search for the products needed to produce the menu through local sources available to foodservice operations in your geographic region. Are all items available? What challenges have arisen in this process? Are there any items that you may need to reevaluate? Why?

Menu for Analysis

- Using the skeleton menu below, fill in each item with a creation that fits with the "Artisan" concept of using fresh, local ingredients and small batch crafted products
- Describe each menu item in an appetizing fashion for your geographic region market
- Provide a recipe for each item, exploded for commercial production for an anticipated 150 diners. If possible, use local purveyors.

Appetizers

Soup of fresh local produce
Shellfish or Chicken in hot or cold form
Regional Ethnic tidbit

Salads

Local Greens
Local Produce

Entrees

Vegetarian made with local products
Poultry made with local ingredients
Pork made with local ingredients
Fish and/or shellfish made with local ingredients
Beef made with local ingredients

Specify vegetable and starch choices as per region

Desserts

Fruit dessert item with local ingredients
Specialty Pastry
Specialty Ice cream/sorbet/Italian ice....with local flair

Beverages

Local wine or liquor unless prohibited
Local caffeinated favorite(s)

Chapter Eleven: Service and the Menu

Objectives

1. Describe the importance of the role of service in fulfilling the objectives of the menu.
2. Explain the concept of service as it relates to the hospitality industry.
3. Identify the essential elements of good service.
4. Differentiate among types of table service, and identify what sort of service might best be matched with various types of menus.
5. Explain the methods in which payment is secured from guests.

Outline

Introduction
- Service is the point at which all work previously done to create the menu comes to fruition
- How patrons are served can be crucial in fulfilling a menu's goals. Poor service can ruin a dining experience. Issues include; timing, temperature of food, replenishment of beverages and accompaniments, full and available condiments, and smooth, effortless fulfillment of needs.

Serving Guests
- Service is an intangible product- not something the guest can "take home"
- The principle of sanctuary arose from the tradition that people must be cared for and not injured while in an establishment. The spirit of hospitality should prevail to provide guests with pleasure and employees with pride and profits.
- Good service requires skill, concentration, knowledge, and diplomacy. It is hard work to constantly hone skills and remain alert to details that affect guests. The payoff is personal satisfaction and increased gratuities.
- Serving staff must have pleasant personalities and enjoy personal interactions with guests and coworkers. They must be motivated, patient and inform management if any guest issues cannot be resolved.
- Servers who solve problems with enthusiasm and grace are true professionals.
- Servers must have stamina and a professional, clean appearance.
- Service will vary with the type and style of the establishment based on the desires of the guests.

  Mise en Place in Service
  - Mise en Place means getting ready for the job to be done so that service can proceed smoothly and efficiently when guests arrive. Chaos does not create a hospitable feeling.
  - Condiments and accompaniments must be readied or assembled
  - Servers must be briefed in preparation and service details at a meeting or by the dining room supervisor.

- The dining area must be clean and in good order. Outdoor areas should also be checked. An eye for detail is important in the dining room.

Greeting Patrons
- The person in host position may greet, seat and take guests to their table and hand out menus. If host is not available, guests should still be made to feel welcome by the first employee they encounter.
- Every guest should feel individually and warmly welcome in a way that is not mechanical. Eye contact helps the guest to feel recognized.

Serving Food
- A good system is needed so that orders will be accurately taken, properly given to the production department, correctly prepared and appropriately served.
- Staff must be trained in standards of service as proscribed in a manual developed for the operation covering conduct and procedures.
- Different foods require different kinds of service. Proper sequence includes timing, and simultaneous arrival of diners' courses and timely arrival of accompaniments including condiments.
- Servers must have a system to recall which guest ordered which food item.
- Firing orders refers to the request for the kitchen to begin making a course. It requires knowledge of preparation time as well as judgment as to the pacing of the guest experience according to their desires.
- Holding spaces for hot and cold food awaiting pick up must hold temperatures
- Expediters may check food so that it is routed correctly to the awaiting guest
- Tableware must be handled so that servers' hands do not touch any part of flatware, glassware or china that will touch the guests' mouth. The use of trays is more professional than using the hands and arms. Carts may also be used though they take up precious space.
- It is still appropriate to serve ladies first, usually proceeding left in a clockwise fashion. Food is served from the left, using the left hand and beverages from the right using the right hand. Service does not have to move in this fashion as long as it is good service.
- It is proper to ask guests if everything is satisfactory once they have had a chance to settle into the meal but not necessary to interrupt. Observation and eye contact are sufficient to perform the check back.
- If the guest has a problem, servers should never argue. A supervisor should be called if the problem cannot be resolved. After service, problems should be addressed so they will not recur.

Type of Service
- The type of service chosen depends on the type of establishment and clientele. Training, having the right equipment and atmosphere are what are required to put good service into effect. Standards ensure proper and consistent service.

Counter Service
- o Fast and usually has a high rate of turnover
- o Counter service does not necessarily save space
- o A server may have 8-20 seats. Everything should be within easy reach
- o Order systems should get orders to the kitchen without having to be physically delivered. Simplification in all things is desired
- o Places may be preset with placemats, sometimes doubling as menus or menus and napkins may be self serve items.
- o Counter service is fast paced and requires excellent organizational skills and quick work habits while maintaining a pleasant manner
- o When not serving, cleaning, replenishing and organizing are all tasks that need to be accomplished. Learning how to keep ahead is essential.
- o After guest has received bill and been thanked, the dishes must be quickly removed and area cleaned for the next diner. Bus tubs should be kept handy and replaced when filled.

Cafeteria Service
- o In cafeteria service, guests go to a counter and select foods that are plated up by service personnel and then take these foods to a table.
- o Modifications include servers taking plates to the table as well as setting up tables with flatware, napkins and water.
- o Good cafeteria service should put about six people per minute through the line
- o A line over 50 feet long is usually not as efficient as a shorter one
- o Cafeteria service is often associated with institutional foodservice or senior citizen markets
- o A scramble system is one in which guests go specific locations to get certain types of food.

Buffet Service
- o Buffet Service has the advantage of reducing service personnel but the disadvantage of wasting food
- o Normally a buffet is served at a set price with guests taking whatever and as much as they want
- o To help control costs, 9" plates, serving staff for higher cost items, using scramble stations and appealing filling starch products are all techniques employed
- o Buffets give simple, fast service for breakfasts, light lunches and for large numbers of people who may wish to eat at different times
- o A smorgasbord is a Swedish buffet that includes a pickled herring, rye bread and mysost or gjetost cheese.
- o A Russian buffet must include caviar in a glass or carved ice bowl, rye bread and sweet butter.
- o Some health food markets and delis have buffets where guests plates or containers are weighed and a per ounce price is charged.
- o Buffets can be combined with other types of service as with salad bars which are included with or supplemental charges to the main meal.
- o Buffets may include carving, waffle, omelet and pasta stations.

- The display of food is important. Varying heights, candles, flowers or more elaborate decorative pieces add appeal. Temperatures and manner of display are regulated by the Health Department.
- A good system of replenishing supplies should be developed.
- Food on a buffet should be the kind that lends to self-service. Thin sauced and hard to handle items should be avoided.

Table Service
- The operation that serves guests at their tables in the familiar way is classified as seated service. This is the basic traditional method
- The three distinct types in the US include; American, Russian, French and on occasion, English service is used.

American Service
- The simplest and least expensive type of seated service is American. It is fast and does not require high labor.
- A silence cloth may be placed under the tablecloth or butcher paper may be taped on the table.
- In American service, food is dished onto plates in the kitchen. The server then takes the plates to the dining area.
- It is common to serve food from the left, using the left hand and beverages from the right using the right. Variation should be based on what is easiest and most efficient to do for the guest
- It is considered most proper not to remove any dirty dishes until everyone at the table is finished with the course. Servers take cues from guests as to their desires. When in doubt, the server must ask the guest
- American service is frequently used at banquets because a large number of guests can be handled quickly.

French Service
- The most elegant, slow and expensive service is French. Historically it involved much ceremony and tasting of many foods.
- Modern service is performed from a gueridon or cart which has a rechaud or heating unit on it. Food may be deboned, portioned or cooked from raw on the cart.
- A maitre d' is in charge of service. In Europe this person exercises much more authority than the American host or hostess counterpart.
- The captain or chef du rang usually supervise about four servers and have charge of a table assisted by the apprentice commis du rang.
- The chef du rang takes the order, gives it to the commis who takes it to the kitchen to the aboyeur who calls them out. When ready, the commis returns the order to the chef du rang who finishes it tableside.
- A sommelier recommends wines, takes the order and serves.
- A service or show plate is placed at the center of the place setting. A hors D'Oeuvre plate may be placed on top of that, if desired.

- The forks are set to the left of the plate and the knives and spoons to the right. The bread and butter plate is above and to the left of the dinner fork.
- The number of courses in French service is limited today.
- The chef du rang must possess showmanship and knowledge of what eating utensils, dishes and methods of service go with which foods.
- Fingerbowls are proper any time needed as well as fresh napkins.
- In formal French service, neither rolls and butter nor salt and pepper are on the table. Only wine and not water is served.
- The Chef du rang presents the check and collects the money. The commis clears.
- The person who comes around with a cart of attractive pastries is the chef du trancheur. Curry persons serve condiments and other food accompaniments.
- French service requires more equipment and space than American or Russian but offers the finest leisurely paced characterized by grace and elegance.

Russian Service
- Russian service has great elegance and showmanship. It is also efficient and relatively fast. It requires less labor and skill than French and is suitable for elaborate banquets.
- In Russian service, the hot or cold plates for the course to follow are put down in front of the guests.
- When all plates are in position, the serving dish is held and each guest is served by the dexterous server.

English Service
- English service is sometimes referred to as formal family service and is used most often in family run inns and on special occasions.
- Foods are brought from the kitchen on platters and in serving dishes.
- The host remains at the table and carves the meat while the hostess serves the vegetables, salad, dessert and beverage.
- The host is served last or next to last before the hostess.
- Servers may carry plates to guests rather than have them passed which requires much handling.
- The first course may be preset on the table and small tables may be placed next to the host and hostess which are later removed.
- Flatware is placed from the outside in, in order to be used.

Organization for Table Service
- The dining room organization will vary considerably based on the different type of operation served.
- Variations on the management of service include; having a head of service, assistant manager acting as head of service, host/ess, head server, or maitre'd.
- The form of organization should be effective and well managed.

- Usually the ratio of hours for service or front of the house to food production or back of the house is 10:7.
  - Person in charge sees that service proceeds properly, servers are neat and clean and established policies are followed.
  - Schedule should be posted in advance with fair assignments
  - Break and meal policy should be established with meals eaten before the shift in conjunction with a shift meeting or during down times

Room Service
- o This may represent revenue and be a significant factor in guest satisfaction
- o Room Service requires considerable space for cart storage, sometimes a separate kitchen or service elevator and special order taking system.
- o Mise en place is very important due to the distance to correct errors or retrieve missing items. It is important to keep hot foods hot and cold foods cold.
- o The check is often time stamped in the kitchen and signed by the guest upon receipt of the meal

Delivery, Drive Through and Self-Service
- o More Americans are working outside the home and less have the skills or inclination to cook daily, therefore, people are eating out more frequently.
- o Delivery allows meal to be brought to the diner. Popular delivery items include pizza and Chinese with steaks, and fine dining also available.
- o Drive through options are generally associated with Quick Service Restaurants. This service often comprises the bulk of the QSR business with some having drive through only and dispensing with indoor dining.
- o Fine Dining operations are countering with curbside options whereby individual or family sized portions available
- o Regardless of whether food is delivered or picked up, order accuracy is critical as guests often do not discover problems until too late to correct resulting in great dissatisfaction
- o Appropriate containers, condiments and disposable ware need to be provided that is both ample and not wasteful
- o Self-service is another way of reducing both labor and food costs for the consumer. Guests must perceive that their efforts translate into greater value.

Handling the Guest Check
- o After guests have finished the server should bring the check upside down and to the right of the host. If the host is not known, it should be placed in the middle of the table. The server should always thank the guests.
- o POS systems make splitting checks simple should guests desire separate checks.
- o A system of maintaining a record of printed checks must be kept though POS systems have eliminated the need where used.
- o POS systems can place orders remotely through a terminal or hand held PDA system.
- o Some operations use expediters who review orders coming in and ensure orders going out are accurate. Appearance and garnish are also checked.

- o A floater check is one that the server holds and reuses for orders and presenting to guests. Computers that read and time stamp checks can reject resubmission of a check or time it out. Collusion still makes it possible to misdirect funds.
- o Checks charged to room service or house accounts must be properly forwarded to the accounting department. Electronic registers again simplify this task.

Gratuities
- o Tips may be added to wage checks or distributed daily. IRS tip reporting records must be maintained.
- o Tipping varies from 10-20% of the total bill with 15-18% being most common.
- o Tips generally make up the bulk of server wages, particularly in states where the hourly minimum for tipped employees is around $2.00.
- o Often establishments add a service charge for parties of six or more or in situations where the diner might not be aware of the custom.
- o While diners should not be forced to tip for bad service, the server should not be denied compensation for work performed by a mean spirited guest.
- o It is common for servers to "tip out" about 15% of their tips to other service staff who assisted in service.

Questions for Review
1. The ancient tradition that people must be cared for and not injured while in an establishment refers to:
   a. Personal service
   b. Sanctuary
   c. Liability
   d. First, do no harm

2. Service varies according the guest's
   a. Expectation
   b. Arriving during a busy or slow period
   c. Luck of the draw with the server
   d. Ability to tip

3. Infrared lights are used to :
   a. Keep cold food cold
   b. Keep hot food hot
   c. Signal servers that their orders are ready to be picked up by the kitchen
   d. Allow guests to signal servers that they have unmet needs

4. What kind of service has high turnover, 8-20 seats with everything within easy reach, and places that may be preset with a placemat doubling as a menu and flatware?
   a. Counter
   b. Cafeteria
   c. Buffet
   d. Table

5. In which type of service do guests go to a counter and select food, plated up by service personnel and than take, or have their food taken, to a table?
   a. Counter
   b. Cafeteria
   c. Buffet
   d. Table

6. Cafeterias are distinguished from buffets in that:
   a. Cafeterias waste more food than buffets
   b. Only cafeterias have scramble stations
   c. Buffets charge according to exactly what guests eat
   d. Cafeterias employ more service personnel

7. The following is the LEAST likely to be suited to buffet service:
   a. Health food markets or delis
   b. Hotel Banquet Departments
   c. Hospitals or Health Care facilities
   d. Hotel complimentary breakfasts

8. Which of the following is NOT a reason to employ someone to carve meat at a buffet?
   a. The danger of guests cutting themselves with a knife
   b. The danger of guests hacking huge hunks of meat, wasting food and destroying the appearance of the roast
   c. The danger of guests being frustrated at this level of self-service, particularly when dressed for a special occasion
   d. To demonstrate that ours is the largest private sector employer in the country

9. The LEAST common type of table service used in the US is:
   a. American
   b. French
   c. Russian
   d. English

10. Wine glasses:
    a. Are set to the upper right of the guest plate
    b. Are heated to speed the "breathing" process
    c. Are chilled to accommodate American palates
    d. Are set directly in front of the guest on the show or service plate

11. Which of the following is NOT a characteristic of formal French service?
    a. Use of fingerbowls
    b. Absence of rolls, butter, salt and pepper on the table
    c. Absence of water and use, instead of wine, as the beverage
    d. Brisk paced service

12. Which scene most likely illustrates English service?
    a. Server flambéing tableside in a sauté pan
    b. Host carving and hostess passing or servers serving plates to guests
    c. Server using a spoon and fork to deftly transfer food from a held plate or platter to the guest plate
    d. A butler-server presenting the guest with a meal of steak and kidney pie from a plate covered with a silver domed lid with flourish and fanfare.

13. The ratio of servers/Front of House to production/Back of House is roughly:
    a. 3:1
    b. 3:2
    c. 2:1
    d. 1:1

14. Delivery, drive-through and self-service are popular due to:
    a. The basic laziness of Americans
    b. Rising fuel prices
    c. Quality of cuisine
    d. People working outside the home with little time, inclination or skills to cook

15. A check that is held by a server to be reused for orders and presented to different guests is a:
    a. Floater
    b. Collusion
    c. Rechaud
    d. Counterfeit

Discussion and Activity Questions

1. What is your view of gratuities? Should they be left entirely to the discretion of the guest or should the establishment set guidelines? Should the back of the house receive a share? Should the front of the house be paid a higher hourly wage or receive a bonus for maintaining food or labor cost goals? Are all guests fair and consistent when it comes to gratuities?
2. Why is it that being a server does not seem to carry the same respect as it does in Europe? What can a manager do to bring dignity and respect to the position? Are Americans more difficult to satisfy as customers than persons from other cultures?
3. Attempt to determine the average amount of gratuity received by each of the following through a survey or informal interviews:
    a. Server in a 24 hour California Menu restaurant such as Waffle House or Perkins.
    b. Server in a fine dining white tablecloth gourmet restaurant.
    c. Server in a banquet department of a hotel or conference center.
    d. A room service server
    e. Delivery driver for pizza, Chinese or other delivered food.
    f. Server in a restaurant that charges a service charge (gratuity included), if you can find such a place.

4. Consider a complaint that a customer may have whiling dining at a restaurant. Analyze the complaint and the server's response and recommend a better way to handle the situation.

Menu Development Activity

- Describe the style of service that best suits your menu(s), what uniform will the servers wear? How will the dining room stations be assigned? What sort of team will serve the tables? (Front/Back Server and Captain with Busser? Self-service, drive-through?) What standard operating procedures will you expect your staff to follow?

- Determine the sort of service that best suits the menu or menus that you have developed. Select a uniform, describe the service system you will use (servers, bus persons, runner? Etc.) to perform service that will best satisfy the needs of your intended market.

Menu for Analysis

See Exhibit 3.10, Deluxe Sunday Brunch menu in Chapter 3. Discuss various service strategies that could be employed with this menu.

Chapter Twelve: The Menu and the Financial Plan

Objectives
1. Explain the need for a menu to cover capital costs and (for commercial operations) contribute to profit.
2. Identify some of the basic costs of going into the foodservice business and indicate how to calculate them.
3. Review some of the methods used to calculate whether a menu is providing an adequate return on investment.
4. Discuss some of the factors that contribute to the success or failure of a restaurant and describe how the positives could be accentuated and negatives avoided.

Outline

Introduction
- Throughout this text, the view of the menu has progressed from a simple view of the menu as a listing, to a more complex definition.
- The menu is the management tool of primary importance for initiating and controlling all work systems in a foodservice operation
- This chapter discusses financial aspects of establishing an establishment and how the menu can be evaluated in terms of satisfying the required return on capital investment.
- Commercial menus must produce enough revenue and institutional menus must stay within reasonable budgetary limits or financial failure will result.
- Operating results can be analyzed to determine whether the menu has met the financial requirements placed on it. Tests are used to measure performance and strategies help managers adjust menus to meet needs.

The Menu and Financial Planning
- Menu dictates facilities, equipment, décor, inventory and space needs. It reflects the essence of the operation

    Capital Investment
    - o A well planned menu can help eliminate or alleviate many capital costs. Specialty equipment must satisfy more than a fraction of the market and should be justified for the cost and floor space.
    - o Minimizing initial capital investment minimizes amount that must be paid back to investors

    Food Service: A High-Risk Business—or Is It?
    - o Banks have labeled our industry high risk- one of the highest of any industry
    - o The Cornell Quarterly published study, Why Restaurants Fail, counters the premise that up to 90% of restaurants fail with figures between 25-30%
    - o Restaurants are subject to risks associated with the complexity of controlling costs, particularly of perishable goods and the variable amount guests are willing to pay.

- It usually takes several investors going together through limited partnerships or small corporations.
- Passive investors contribute only capital while in other situations; expertise is lent along with monies. Regardless, some form of borrowing must take place as well.
- $500,000-over $1M may be required in capital investment with lenders requiring investors to put up 30% or more
- Franchises may be purchased for as little as $10,000 but frequently requires considerably more to yield higher returns.
- The complex tasks of merchandising, food purchasing, production and service, as well as finance will need to be performed by a few experts or an extremely talented few individuals.
- This may be one reason restaurants fail. Competition and the failure to understand the complexity of the undertaking has been the downfall of many.

The Feasibility Study
- The feasibility study is a detailed report to show exactly what the capital investment is to be used or and precisely how the operators plan on generating the revenues to pay back that investment.
- Independent contractors will work with an operator to write a feasibility study. They can be expensive but also provide assistance with obtaining funds from lenders.
- Information included in the study consists of: the concept, potential market, what it will cost and what kind of return can be expected. A formal study may be hundreds of pages long. The operator must not be passive but understand and approve all aspects of the study.

Restaurant Business Plan Software Packages
- This is an alternative to a professionally done study
- Software varies in quality and usefulness so smart shopping is wise
- User must do a good deal of information gathering which forces the operator to realize the extent of planning and detail that must be worked through to hope to achieve success

Analyzing the Feasibility Study
- Data gathering is the first and most time consuming stage of the process
- Residential and commercial demand is measured by area demographics, traffic patterns, and planned development however the Cornell study stated that having a defined target market or marketing strategy is less critical than public relations, community involvement and customer relations.
- Concept development step involves the use of the population data to form a concept that meets their needs and preferences. Elements of concept include: theme, menu, service, style, hours, and atmosphere.
- The Cornell study cited a well defined concept as the first essential component of success. Losing focus and trying to be too many things to too many people were reasons for failures and closure.

The Short-form Feasibility Study
- While no substitute for an in depth study, it can help to fine tune the concept prior to a formal study.

Location and the Market
- Location alone may be decisive in determining whether the operation even has a market
- Site analysis is the stage of the process whereby it is determined if the selected site could support a new restaurant concept.
- It is important to know how many potential customers come by the door and what method of travel that they use and whether they might come.
- An operation must be suited to the kind of traffic it has and parking available
- The location should be investigated for environmental factors. The Cornell study found that a poor location can be overcome by a great product and operation but a good location cannot overcome bad product operation.
- Competitor analysis is the survey of area restaurants, delivery, food market, take out, and quality prepared groceries. Some operations may be complimentary..
- 2005 NRA Table Service Trends Report cited competition as the biggest challenge their business will face though the Cornell study stated that successful owners viewed competition to be tool for self measurement only.
- What appears to be an existing market may not be durable. There is a tendency to overestimate rather than underestimate, markets
- Numbers of potential market members must be supplemented with a market study asking who they are and if they can be well defined.
- Defining characteristics include: income, social habits, eating patterns, ethnic background, age, sex, and others.
- What do they want and can it be provided? Why will they come, where do they eat now and what do they eat? Is there competition and is the market steady and reliable?

Own, Rent, or Lease?
- There are a number of ways the operation can be established which will effect costs
- A new building can be erected and furnished, belonging to the owner, the land can be leased with the landowner putting up the building, or the site and building can be built by someone else and leased.
- Ways to arrange to rent or lease are various
- A straight rental of so much per year, month or time period may be negotiated based on a percentage of gross sales.
- The highest rent would be for a turnkey operation or grade 4 lease with everything ready to go.

- A net net net lease is one in which the business owner pays rent, taxes and insurance. In a net net lease the operator pays only rent and taxes. A net lease means that only rent is paid however the business owner must usually carry the insurance on the equipment and furnishings heor she owns.
- The length of the lease is variable. Due to the instability of many foodservices, leases often run for five years and less
- A grade 3 lease covers an improved building shell with food preparation equipment in it. Rent is lower than for grade 4 (turnkey operation)
- A grade 2 lease is for a slightly improved shell including partitions, finished floor, finished coiling, heating ventilation, electrical wiring, and outlets, air conditioning, plumbing with fixtures, and drains. Rental may run 6-8% of gross sales.
- The lowest level rental arrangement is a grade 1 lease which is an unimproved shell
- Variable rent is usually based on a percentage of gross sales or a minimum rent is paid plus a percentage or a sliding percentage of gross sales with the percent figure declining as sales increase
- In negotiating a lease it is important to establish exactly when it is to start. If rent is paid during remodeling or building before opening, the property may have to be operated for some time to erase the rental deficit and show a profit.

Other Capital Costs

- Pre-opening expenses are funds spent prior to opening for non-assets. They include salaries paid prior to opening, marketing, licenses and other prepaid expenses, utility deposits, and other costs incurred prior to the opening.
- Working capital is the money necessary to keep the operation afloat once it is in business. It includes cash reserves for emergencies, funds needed to buy supplies, inventory at quantity discounts, and money tied up in cashiers' banks.
- Pre-opening expenses are often overwhelming. The range of costs can total over half a million dollars. They include; attorneys and accountants fees, permits licenses and fees collected by government, insurance, utility deposits, advertising, and menu printer.
- Stocking with inventories can be quite expensive. Average food inventory turns over about once every two weeks
- Inventories are seen variously as checking accounts where funds move in and out quickly, or as savings accounts where funds sit and gain interest of volume purchasing.
- Liquor inventories are said to turn about eight times a year. Many keep tow- one for wines and one for beers and spirits. Beers and spirits should turn over about every two weeks. Liquor can be bought at discount cases prices
- One final pre-opening expense is the hiring of staff as skilled employees should already be trained and on the payroll when the business opens.
- Once the doors open, working capital in the form of cash for cash registers and money to buy more food, liquor and incidentals is needed. There should be one half normal monthly expenses in the checking account.

- Additionally, the operation should keep a contingency fund for emergencies. This may be replaced with a borrowing agreement with the banker

Financial Analysis and the Menu
- Financial success is measured by profitability and return on investment.

Measuring Profitability
- The return on assets is the most common measure of how well a business is performing. It shows how well the assets have been used. 3-6% is considered a fair return.
- Net profit is the profit left over after operations, interest, other non-operating expenses, and business taxes have been paid.

Return on Investment
- Return on Investment is the amount of money obtained as a profit in relation to the actual amount invested by the owners.
- Investors usually demand that the ROI be higher than that on relatively safer investments such as government bonds
- Leverage is the amount invested versus the asset value. If investors can get high financing they can often buy an operation that generates high profits on relatively little investment. Finding lenders willing to take that risk is difficult

Cash Flow
- Cash flow is the dollars generated in the operation during the operating period-dollars that can be used to pay off obligations plus other costs
- Sources of cash flow include: sales revenue, additional invested cash, or net proceeds from additional loans. Cash flow is calculated by deducting payments made to banks to pay principal on loans and other cash uses not ordinarily recorded on profit and loss statements
- Operations may do well on paper due to accounting conventions but poor managerial decisions may leave little money to pay creditors when bills are due. Dollars, not paper pay bills.
- Management should know about cash flow and impact on future cash flows before making capital investment decisions. No investment is sound if the cash flows generated after the investment are insufficient to repay the loans and, ideally, additional cash to the owners

Liquidity Factors in Food Service
- Liquidity means that current assets such as accounts receivable, inventory, cash and other convertible assets, exceed liabilities or debts such as capital payments due within the current accounting period and accounts payable.
- Current ratio is used to determine liquidity and is calculated by dividing current assets by current liabilities. Assets should be greater than liability through 1:1 is considered barely adequate, 2:1 or 3:1 is better.
- Another calculation method is the ratio between dollar sales and working capital. Working capital is current assets less current liabilities. A high ratio such as $40 or $50:1 is desirable.

- The solvency ratio indicates worth in respect to debt. This is the ratio of owners equity to the amount owed to creditors. 1:1 is adequate but creditors prefer 2:1.

Noncommercial Operations
- Measuring the operating efficiency of a noncommercial operation is often simpler than for a commercial one. The most common check is to compare operating costs and sales to budgeted figures.

Menu Strategies and Financial Success
- Menu prices may be changed to increase check averages if menu items are changed or have perceived value to match
- Prices could be lowered and volume increased
- A business could add delivery
- The feasibility study should not be thought of as fixed but a dynamic document with allowances for adjustment for opportunities.
- Success is not measured in dollars alone but the pleasure the operation brings to the guests and thus the owners.

Questions for Review

1. To be successful, a restaurant operator must:
    a. Enjoy hosting dinner parties
    b. Be in possession of a real winner of a recipe
    c. Be ready to relax and enjoy a leisurely paced lifestyle
    d. Be prepared to work long hours, be actively involved in the day to day operations and practice diplomacy.
2. The advantages of using Business Plan Software (versus having a feasibility study done by professional consultants) include all of the following EXCEPT:
    a. The software performs the task of data collection for the user
    b. It may save money
    c. By doing a little homework, an operator can purchase software of high quality and usefulness
    d. It gives the prospective operator a taste of just how much work and detail goes into planning.
3. The 2005 National Restaurant Association Table Service Trends Report cites what? As the biggest challenge their business will face:
    a. Immigration Reform
    b. Competition
    c. Minimum Wage
    d. Tip Reporting
4. Various ways to arrange to rent or lease a building for a foodservice operation include all of the following EXCEPT:
    a. Straight rental by the month or year
    b. Net net net lease
    c. Sliding Scale where the percentage increases as sales increase
    d. Variable rent based on gross sales

5. A grade 3 lease is:
   a. An improved building shell with food preparation equipment in it
   b. A turnkey operation
   c. An unimproved shell
   d. A slightly improved shell
6. The following statement is true about pre-opening expenses:
   a. They are minimal
   b. They are often overwhelming
   c. They are part of working capital
   d. They are covered by cash reserves
7. Average Food inventory turnover is about:
   a. 12 times a year
   b. 26 times a year
   c. 52 times a year
   d. 365 times a year
8. Liquor inventories turnover about:
   a. 4 times a year
   b. 8 times a year
   c. 12 times a year
   d. 52 times a year
9. Working capital is used for all of the following EXCEPT:
   a. Cash for cash reserves
   b. Money for supplies; food, liquor, and incidentals as needed
   c. Checking account; the balance recommended is to equal 50% of normal monthly expenses
   d. Capital expenses
10. If assets are $750,000 and profits are $38,000, the ratio of profit to asses is:
    a. 5%
    b. 19.74%
    c. 50.7%
    d. 71%
11. If the owner invests $750,000 and makes a profit of 80,000, then the ROI equals:
    a. 935.5%
    b. $830,000
    c. A loss of $670,000
    d. 8.52%
12. Depreciation and accounts payable can distort the true perception of:
    a. What cash is really available to operators to pay bills when due
    b. Profitability
    c. Accountability
    d. Solvency
13. Current liabilities include all EXCEPT:
    a. Accounts payable
    b. Mortgage payable
    c. Inventory
    d. Payroll

14. The solvency ratio preferred by creditors is:
    a. 1:1
    b. 1:2
    c. 1:4
    d. 2:1
15. The Foodservice industry is a risky one. Risks typically cited include all of the following EXCEPT:
    a. High capital to open and run a business
    b. The perishability of the products
    c. High cost of employee benefits
    d. The labor intensive nature of the business

Discussion and Activity Questions

1. How many members of the class dream of owning their own restaurant someday? Is the information presented in this chapter sobering or encouraging? Discuss.
2. Interview a local, successful entrepreneur and ask him for the secrets of his success. (Or invite one to come be a guest speaker)
3. What is so alluring about the restaurant industry for those who have little or no experience that people are willing to invest (and lose) large sums of money in a venture? Why are you drawn to this sometimes risky business?

Menu Development Activity

Do an abbreviated feasibility study for the menu(s) that you have developed over the course of the semester. Does it (do they) represent feasible possibilities? Are there alterations that you could make that would make the menu viable if it does not seem so now? Make any changes that you have discovered may be necessary to successfully operate your menu.

Menu for Analysis

Turn back to the original RJ Grunts Menu, exhibit 1.3 in Chapter 1. This is the menu that launched an extremely successful restaurant empire, Lettuce Entertain You Enterprises. This is a menu from decades ago. Other than the whimsically, for now, low prices, what is appealing about the menu? What kind of mood would you expect to find at RJ Grunts? Does the menu convey food that tastes good? How? What else are you getting from the menu? Is there anything you would change about it? Why? What would you order?

If you are unable to clearly read the menu, go the Lettuce Entertain You website and download the current RJ Grunts menu and answer the same questions, albeit knowing it is no longer the original menu you are talking about.

Chapter Thirteen: Ethical Leadership in Restaurant Management

Objectives

1. Define the qualities of leadership that contribute to the successful operation of a foodservice establishment.
2. Explain the elements of effective communication in the workplace
3. Outline methods by which various sources of workforce motivation can be discovered, nurtured and maintained.
4. Identify ways a dynamic approach to problem solving helps to avoid crises and the need for crisis management.
5. Define the concept of business ethics and elaborate on the relationship between leadership and ethics.
6. Discuss the ethical responsibility of the foodservice industry with regard to providing nutritional value to its patrons.
7. Discuss the ethical responsibility that foodservice operators have in terms of their membership in the foodservice industry and community.

Outline

Introduction

- No matter how much work has gone before, a foodservice enterprise will not be successful without Leadership
- Every operation has a manager but not every manager is a leader
- Managers operate the position of being in charge but may not have the respect of the crew
- Managers plan and organize mechanically, reacting to events, pushing products and controlling people
- Leaders envision the future, followed by a staff that looks up to them
- Leaders pull guests toward quality offerings and provide a motivating climate for people while controlling things
- Managers may suffer from burnout while Leaders look forward to going to work

Leadership in the Foodservice Industry

- Business tasks and employee motivation needs must be blended in today's changing world
- Businesses do not compete simply for guests, but for quality staff as well
- Leaders establish standards that create and maintain a quality group of employees
- Leaders build the team that desire to work to accomplish the common goals of the organization
- A manager becomes a Leader through respect based on: vision, character, abilities, knowledge, and professional accomplishments. Employees respond to being treated with dignity and respect.

Leadership and Self-Awareness
- Successful leaders embody integrity, enthusiasm and self-awareness.
- Leaders set an example for employees both professionally and personally
- Prioritizing, delegation, time management, communication, team building and problem solving are all crucial skills with a healthy sense of humor
- Relaxing and enjoying a personal life outside of a five-day workweek contribute to the ability to Lead
- Drugs and alcohol are problems that are so prevalent in the industry that illicit usage was reported to be higher than any other. Managers must set an example and be mindful of employee behavior that affects job performance.

Communication and Leadership
- Effective communication is crucial to success both with guests and employees
- Barriers to effective listening include:
  - Language- sender and listener must use words that both understand mean the same thing
  - Culture- sender and listener may share culture or be more diverse. Bias, prejudice or assumptions are barriers as well as style.
  - Environment- sender and listener both need a lack of distractions such as noise, pressing demands, lack of privacy or poor timing
  - Assumptions- sender and listener must feel free to ask clarifying questions, seek feedback and interact honestly yet, tactfully.

- Communications in foodservice is typically verbal but written communication is also used and is important
  - Writing must have a purpose and be planned, clear, succinct and in an accessible style
  - Always proofread and be aware of tone; sarcastic humor and an accusatory tone puts employees on the defensive
  - Too much detail can bog down the communication
  - Many adults are illiterate and may not admit this readily. Adult literacy courses can be a much appreciated benefit

- Meetings are opportunities for Group Communication
  - Question and Answer sessions, solicitation of feedback, and generation of enthusiasm are all possible
  - Meetings need to remain focused on resolving issues and achieving communication goals
  - Expectation of respectful cooperative communication between front and back of the house set a hospitable tone and create positive work climate
  - Be clear about the nature of the meeting so employees know what to expect. Informational announcement meetings should not be billed as something where input is sought.
  - Employees should be paid for mandatory meetings to remain ethical and to achieve their receptive presence

Motivation and the Hospitality Leader
- Motivation comes from within, according to theory, therefore the manager must know what motivates each person by getting to know them
- Once known, these factors must be nurtured and maintained, with the recognition that what motivates people may change over time
- Treating employees with dignity and respect is a simple universal principle which eliminates sexual harassment, discrimination and provides a guideline for addressing complaints and conflicts.
- An appreciation for the diverse nature of humans assists the leader in acknowledging and accepting all
- Employees who feel dignified and respected are better able to serve guests and are more likely to be motivated and satisfied in their jobs.

Setting Service Charges versus Gratuities
- The traditional practice of guests tipping servers is under discussion
- Negatives include; reducing the server to "servant" stature with the guest who may or may not tip due to poor service or matters completely unrelated to the quality of the job performed, and putting the server in the position of putting up with bad behavior from the guest in fear of retaliation
- Service charges have been used in Europe whereby an automatic gratuity is added to the bill. This amount can be disputed just as any other item which is not up to standard but provides an equitable amount to base pay and taxes on.
- A service charge system may help ease some of the traditional friction between the front and back of the house as servers are not penalized for actions of the kitchen that are beyond their control
- Traditionalists will want to keep the current system which has provided the lion's share of most servers' incomes.

Creating Community in the Workplace
- Once the individuals are known, the Leader can focus on teambuilding
- Creating community helps to encourage an atmosphere where employees and employers do not dread coming to work
- Self-care, communication, motivation, and treating employees with dignity and respect are the building blocks.
- Quality produces pride which helps build teams
- The leader sets the tone for transactions with guests, employees, purveyors and neighbors
- Equitable systems for scheduling, work assignments, promotions, raises and discipline create a climate of quality
- A pay system where everyone has a stake in the success of the operation builds teams
- Cross-training helps employees see the big picture and helps with scheduling
- Healthy extracurricular activities can be positive

Problem Solving

- Processes that solve problems help avoid crisis and discourage burnout that results from constantly putting out fires
- Soliciting feedback with a sincere desire to act on information can be achieved through managers seeking guest input, comment cards, websites, toll free numbers and mystery shoppers
- Employees are another source for information about how to improve
- Policies should focus on guest satisfaction
- Jobs should be designed so that they can be reasonably performed by a well trained person without a feeling of chaos
- The process method for problem solving consists of these steps:
    1. Define the problem
    2. Determine the root cause
    3. Determine alternative solutions and the consequences
    4. Do the benefits outweigh the costs?
    5. Is it sound?
    6. Will it fix the system flaws?
    7. Will it work?
- Problems that are not solved may worsen and manifest into crises
- Decreased morale and resulting turnover and loss of profits are problems for managers who are not leaders and ignore problems
- Disaster and emergency procedures that are established can provide guidelines to keep a bad situation from spinning out of control

Ethics

- Personal ethics represent some form of the golden rule, the precept of all the major world religions
- Business ethics are those behaviors governing the way supervisors and employees conduct business
- Ten Ethical principles for Hospitality Managers were adapted from the Josephson Institute of Ethics "Core Ethical Principles"
    1. Honesty
    2. Integrity
    3. Trustworthiness
    4. Loyalty
    5. Fairness
    6. Concern for and respect for others
    7. Commitment to Excellence
    8. Leadership
    9. Reputation and Morale
    10. Accountability
- These ten principles serve as guidelines for making decisions and about how to behave

Hospitality Ethical Issues

Nutrition

- Historically, responsibility for nutrition was considered to be proportional to guest choices for dining. Those with little or no alternatives such as prisoners, patients and students were considered to be those that foodservice had a duty to provide a high quality of nutritional worth
- Guests with more freedom historically dined away from home primarily for special occasions, and indulged themselves infrequently. There did not seem to be a strong duty to provide nutritional worth
- Americans are less active and more likely to experience stress for which food is used to soothe
- The Keystone Report of the FDA indicates that 64% of Americans are overweight, including the 30% defined as obese.
- The report suggests a connection with the rise in obesity with the increased incidence of consuming food outside the home in restaurants
- Consumer advocate groups have requested better labeling, reduction in portion sizes, and an increase in healthy choices that are appealing
- The quick service segment is perceived to be a particular culprit, abetted by the films Super Size Me and Fast Food Nation.
- One proposed solution is to introduce flavorful food items that happen to be healthy; vibrant flavors, vegetarian options and an emphasis on healthy fats and whole grains
- Healthydiningfinder.com is a website linked through the NRA that helps restaurants identify healthy options on their menus and assists with product development as well as nutritional analysis

Labor
- The food service industry is often the first rung on the ladder to the American dream. It is a strong employer of newly arrived immigrants
- The NRA provides a link to Daily Dose Language Systems Inc. which offers job specific English programs for the workplace

Piece of the Success Pie
- Costs and causes of turnover have been examined and one cause has been thought to be the lack of stake that many employees, including managers, have in the operation
- Opportunities to become managing partners with partial ownership in the "store" are designed to increase loyalty

- Group Health Care plans, tuition reimbursement, and other benefit options decrease turnover and increase stability, loyalty and motivation
- Providing realistic means to achieve bonuses is also motivating

Good Neighbor

- Professionals who work together toward the betterment of our industry and communities provide a network. They include: NRA and their Educational Foundation, ACF, FENI and CHRIE
- Education, mentorship and other forms of resources can be found through these organizations

Keeping it Green

- The Green Restaurant Association is a national environmental organization founded in 1990 to help restaurants and their customers become more environmentally aware and sustainable.
- The Five GRA components are:
    1. Research
    2. Environmental Consulting
    3. Education
    4. Public Relations and Marketing
    5. Community Organizing and Consumer Activism

Charitable Contributions

- The foodservice industry prides itself on disaster response through the provision of food to rescue workers and those in need of rescue.
- Perhaps some of the same motivators that lead people to the hospitality industry evokes a hospitable, charitable response to hunger
- Locally, charitable and civic food donations, leftover pick up services from shelters, and market related altruism can make a difference in the communities in which foodservice operates
- The National Restaurant Association offers guidelines for giving

Questions for Review

1. The respect builders that can change a "manager" to a "leader" include all of the following EXCEPT:
    a. Vision that others believe in and follow
    b. Professionalism
    c. Fear of losing the job
    d. Character
2. In order to communicate effectively, written documents should always:
    a. Provide lengthy, detailed descriptions or instructions
    b. Lighten up with some tongue in cheek humor
    c. Assume that readers are operating at a High School reading level
    d. Be clear, succinct and planned

3. Meetings are an opportunity for all of the following EXCEPT:
    a. Announcements about policy changes or problems with no input allowed
    b. Question and Answer Session
    c. Solicitation of feedback
    d. Information regarding forecasts and specials
4. Generally, people work for all of the following reasons EXCEPT:
    a. Provide for their families
    b. Occupy themselves
    c. Opportunity for Leadership
    d. Satisfaction of contribution to the success of an operation that they are proud of
5. The following are all positive aspects of the gratuity system for servers EXCEPT:
    a. Traditional and expected
    b. Gives the guest a feeling of control
    c. Modeled on the European service charge system
    d. Motivates servers to "hustle" for the guest
6. Feedback from guests can be effectively solicited by all of the following EXCEPT:
    a. Managers sincere desire to enhance guest enjoyment
    b. Comment cards, websites and toll free phone numbers set up for the purpose of gathering guest response
    c. Mystery shoppers
    d. Routinely performed "check back"
7. The first step in solving a problem is to:
    a. Define it.
    b. Develop a plan
    c. Brainstorm
    d. Document the process
8. When determining alternative solutions, in the problem solving process, ALL of the following should be asked EXCEPT:
    a. Do the benefits outweigh the costs?
    b. Will it work?
    c. What is the cheapest solution offered?
    d. What are the consequences?
9. Those behaviors governing the way supervisors or other representatives of the operation conduct business is known as:
    a. The Golden Rule
    b. Business Ethnics
    c. Peter Principle
    d. Equal Opportunity Employment
10. Which of the Ten Ethical Principles for Hospitality Managers would be helpful in analyzing the values embodied in Truth in Menu?
    a. None- that is a legal, not an ethical matter
    b. All of them
    c. Honesty
    d. Honesty, Integrity, Trustworthiness, Commitment to Excellence, Leadership, Reputation and Morale, and Accountability

11. According to the Keystone Report, funded by the Food and Drug Administration, the percentage of overweight adults in the United States is almost:
   a. 25%
   b. 45%
   c. 65%
   d. 85%

12. All of the following are considered to help reduce turnover EXCEPT:
   a. Having a stake in the operation through owning a piece of the "store"
   b. Group Health Care Benefits
   c. Tuition Reimbursement Programs
   d. Bonuses that remain just out of reach, thus providing motivation

13. The GRA founded in 1990 stands for:
   a. Green Restaurant Association
   b. Great Restaurants of America
   c. Giving Restaurateurs Advantages
   d. Grain Reduction Act

14. Which is NOT a professional association advocating professionalism in the foodservice industry?
   a. CHRIE; Council on Hotel, Restaurant and Institutional Education
   b. FDA; Food and Drug Administration
   c. FENI; Foodservice Educators Network International
   d. ACF; American Culinary Federation

15. Now that you have completed the course and your textbook, what role does the menu play in Restaurant Management?
   a. It is the central document at the core of the successful operation of the foodservice enterprise
   b. Very little.
   c. It is the finishing touch, once everything else has been put in place.
   d. It is a job for the chef

Discussion and Activity Questions

1. What is the difference between a "manager" and a "leader"?
2. Give an example of behavior that you have experienced from a supervisor that you would classify as "manager" behavior? "Leader" behavior?
3. Describe an ethical dilemma you have encountered in your working life or that you have observed. Could using the Ten Ethical Principles for Hospitality Managers as a guide have assisted in solving or at least defining the issue? How?
4. What are your views on liability for our industry versus personal responsibility regarding the following issues. Explain your views.
   a. Alcohol and liquor liability
   b. Obesity
   c. Sexual Harassment
5. Describe an effective team building activity or technique employed that you have observed or experienced in the hospitality industry. Describe a management activity or technique that served to pit employees against each other. Which were you able to find more examples of?

Menu Development Activity

Now that you have completed your menu project, describe any challenges you might have faced in the execution of the menu. Whether or not you actually performed a function with your menu, describe your leadership strengths and areas for growth. What self-development activities can you engage in to help you increase your ability to lead others to their top performance?

Menu for Analysis

Evaluate the menu that appears below for its contribution to ethical management. Are there any ethical problems created by Tubby's menu? What changes could be made to solve any ethical dilemmas that the menu poses? Are there any other problems with the menu?

Tubby's

Teaser's

Fried Cheese with Mayonnaise Dipping Sauce

Fried Chicken Strips with Mayonnaise Dipping Sauce

Fried Onions and French Fries Combo topped with rainforest beef chili and cheese

Belly Busters
Served with fried pickles and hush puppies

Pulled Pork
Raised on a local industrial farm which has gone 20 days without a lagoon spill. The latest technology is keeping this meat free from vermin.

Chicken Licken
Deep fried and quick raised on steroids to produce the biggest breasts and tastiest thighs around. These birds just might compete with our servers, Tubby's Chickies. So good your gonna crow for your supper!

Chicken Fried Steak
Pounded Rainforest Beef is battered and fried to golden and served with sausage cream gravy. 12 ounce portion is sure to leave room for dessert.

Fried Oysters
Deep fried mollusks from restricted waters provide that special wallop with our own tartar sauce- Fresh from Uncle Jimmy's boat-

Eggs Benedict
Country Ham over white flour biscuit made with old fashioned Lard is ladled with hollandaise sauce for that late night appetite or breakfast anytime. Tubby's Serves only double yolked eggs with fresh churned Margarine

Sweet Nothings

Giant Chocolate Chip Cookie Marshmallow Sandwich

Bubble Gum Ice Cream with Maraschino Cherry Topping

Cotton Candy- Spun sugar is spun with artificial color (Free with any entrée from our Kid's meal)

Whistle Wetters

Soda Pop- every artificial flavor you can think of with a caffeine jolt- bottled in a third world country by malnourished children

Shake- dairy flavored soft serve blended with every artificial flavor you can think of and some you can't. Ask us about mixing in candy bar bits, smashed cookies or bubble gum

Beer- all you can drink for a penny a pound of your weight. Add free shots with every pitcher during our wet t-shirt contest

Any item can be prepared to-go in our indestructible Styrofoam containers

Answer Key to Questions for Review

Chapter 1
Questions for Review
1. C
2. B
3. D
4. A
5. C
6. B
7. D
8. D
9. A
10. C
11. B
12. B
13. C
14. B
15. A

Chapter 2
Questions for Review
1. C
2. D
3. C
4. A
5. A
6. B
7. A
8. B
9. D
10. C
11. D
12. C
13. B
14. B
15. C

Chapter 3
Questions for Review Answer Key
1. D
2. A
3. B
4. B
5. D
6. C
7. D
8. C
9. B
10. A
11. C
12. C
13. A
14. B
15. D

Chapter 4
Questions for Review
1. C
2. B
3. C
4. A
5. B
6. B
7. D
8. A
9. D
10. A

Chapter 5
Questions for Review
1. B
2. A
3. D
4. C
5. B
6. D
7. C
8. A
9. C
10. A

Chapter 6
Questions for Review
1. B
2. C
3. A
4. C
5. B
6. B
7. C
8. A
9. D
10. A

Chapter 7
Questions for Review
1. B
2. D
3. B
4. B
5. D
6. A
7. C
8. B
9. A
10. D

Chapter 8
Questions for Review
1. C
2. D
3. A
4. A
5. B
6. C
7. A
8. B
9. C
10. A

Chapter 9
Questions for Review
1. B
2. B
3. A
4. D
5. C
6. C
7. A
8. B
9. B
10. D

Chapter 10
Questions for Review
1. B
2. D
3. A
4. C
5. D
6. A
7. B
8. C
9. C
10. B

Chapter 11
Questions for Review
1. B
2. A
3. B
4. A
5. B
6. D
7. C
8. D
9. D
10. A
11. D
12. B
13. B
14. D
15. A

Chapter 12
Questions for Review

1. D
2. A
3. B
4. C
5. A
6. B
7. B
8. B
9. D
10. A
11. D
12. A
13. C
14. D
15. C

Chapter 13
Questions for Review

1. C
2. D
3. A
4. B
5. C
6. D
7. A
8. C
9. B
10. D
11. C
12. D
13. A
14. B
15. A